Study Topics in Physics **Book 3**

# Energy

W Bolton

SHELL/ESSO
BRENT C

**Butterworths**
London   Boston
Sydney   Wellington   Durban   Toronto

| United Kingdom | **Butterworth & Co (Publishers) Ltd** |
| London | 88 Kingsway, WC2B 6AB |
| **Australia** | **Butterworths Pty Ltd** |
| Sydney | 586 Pacific Highway, Chatswood, NSW 2067 |
| | Also at Melbourne, Brisbane, Adelaide and Perth |
| **Canada** | **Butterworth & Co (Canada) Ltd** |
| Toronto | 2265 Midland Avenue, Scarborough, Ontario, M1P 4S1 |
| **New Zealand** | **Butterworths of New Zealand Ltd** |
| Wellington | T & W Young Building, 77-85 Customhouse Quay, 1, CPO Box 472 |
| **South Africa** | **Butterworth & Co (South Africa) (Pty) Ltd** |
| Durban | 152-154 Gale Street |
| **USA** | **Butterworth (Publishers) Inc** |
| Boston | 10 Tower Office Park, Woburn, Massachusetts 01801 |

First published 1980

© Butterworth & Co (Publishers) Ltd, 1980

ISBN 0 408 10654 9

**British Library Cataloguing in Publication Data**

*Bolton*, William
    Energy. — (*Study topics in physics; book 3*).
    1. *Force and energy*
    I. Title    II. Series
    531'.6      QC73      80-40009

    ISBN 0-408-10654-9

Typeset by Scribe Design, Gillingham, Kent
Printed and bound by Whitefriars Press Ltd.,
Tonbridge & London

# Preface

This series of books has been designed to cover the main aspects of physics courses that, in the U.K. at least, are generally taken by students aged sixteen to eighteen years, often prior to going to university. The series, while based upon the requirements of modern 'A'-level syllabuses, reflects the shift of emphasis in the teaching of physics in recent years and maintains a careful balance between the best of traditional courses and recent innovations. By selecting books from the series, and possibly supplementing with additional specialist material, a wide variety of courses can be covered. Some books, for example, could well be used in some of the technician courses in colleges.

Each book in the series is designed to cover a main topic in physics and each has been written in a feasible teaching sequence which carefully develops the structure of the physics. The chapters have been written in an essentially self-teaching method with text and questions interwoven. All the questions used within the text are supplied with suggestions for answers. At the end of each chapter are further questions, no answers supplied, which could be used for assessment. Many chapters also include background reading.

Though the physics in the books is developed from a basis of experimental data, details of experiments are not included. This is to enable teachers to use the book with the apparatus they have available and so plan the experimental work to suit their resources. References to sources of experiments are given.

The books have undoubtedly been influenced by my earlier work with the Nuffield Foundation Advanced Physics Teaching Project and UNESCO, as well as my present work with the Technician Education Council. The influence of the Physical Science Study Committee (PSSC) course in physics and the Project Physics Course is also apparent. The form of the books and the way the physics is presented is, however, my interpretation of the subject and any errors mine.

W. Bolton

# Contents

# 1 The concept of energy

**Objectives**

The intention of this chapter is to probe, and possibly enable you to revise, the concept of energy in relation to its conservation. Knowledge of the concepts associated with motion, Newton's laws, elasticity and rotational motion are assumed. The chapter could ideally follow on from the material in *Book 1: Motion and Force* and *Book 2: Materials.* A different sequence might mean the omission of some of the sections in this chapter.

The general objectives for this chapter are that after working through it you should be able to:

(a) Explain what is meant by energy;

(b) Explain what is meant by the conservation of energy;

(c) Define kinetic and potential energy;

(d) Define work and power and solve problems involving them;

(e) Carry out calculations involving kinetic energy, potential energy and work;

(f) Explain what is meant by gravitational potential energy, elastic potential energy and strain energy and carry out calculations involving them;

(g) Define rotational kinetic energy and carry out calculations involving it and work for rotating objects;

(h) Explain energy transformations involving energy forms such as potential energy, kinetic energy, strain energy, electrical energy, work and heat;

(i) State the first law of thermodynamics and define the term internal energy.

*Loch Awe pumped storage scheme*

**Teaching note**

Experiments appropriate to this chapter can be found in *Nuffield Physics: Pupils' Text Year 4, Physical Science Study Committee Laboratory Guide for Physics* and *Project Physics Course Handbook.* (Full bibliographic details are given on p. 94.)

## What is energy?

You probably already have some concept of energy. A simple concept is to consider it to be the 'go' of things. If an engine is supplied with energy, perhaps obtained from steam under pressure or exploding petrol, it 'goes', i.e., it runs and can do jobs. A motor could be supplied with electricity and 'go', i.e., revolve and perhaps lift objects – do jobs. If you eat your cornflakes you, a human engine, receive energy and are able to do jobs. In all these different situations, and many others even more diverse, something is involved which we call **energy**.

But why have this concept of energy? How are we able to reconcile the different forms of it – they are so diverse, how are we able to give them all the same name of energy? Energy is something that is conserved. That is, if at any instant we add up all the pieces of energy, in whatever form, and arrive at a number and then, at some later time, when the energy pieces may be in different forms as a result of changes, we again add up the pieces we will arrive at the same number as we started with. Whenever we add up the pieces of energy in a system we always reach the same number. We devise formulae for the energy in its different forms so that whenever we add them up they always come to the same number. This principle is known as the **conservation of energy**.

Conservation laws are very powerful laws and yet rather simple laws. They apply to all objects, large or small, and wherever they are. The laws are simple because they can be stated as – something remains constant and never changes.

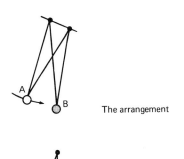

The arrangement

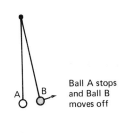

Ball A pulled to one side, and released, Ball B at rest

**Question** **1** In Dalton's theory of atoms in chemical reactions atoms were considered to be conserved, the number of atoms remaining unchanged despite changes in arrangements. In chemical reactions, atoms are neither created nor destroyed, but only rearranged. Explain this conservation law by means of an example.
(*Note*: this is not a good conservation law because it is not always applicable – matter can be destroyed and converted to energy; the sum of mass and energy is conserved.)

It is important to realize that the principle of the conservation of energy does not tell us whether a change will take place but only that if it does there will be no change in the total energy. The same is true of Dalton's theory of atoms in chemical reactions – you cannot tell whether a reaction will occur but only that if it does there will be only a rearrangement of the atoms.

## Kinetic energy – conservation

*Figure 1.1* shows a simple piece of apparatus in which one ball, often made of steel, swings down and hits another ball that is initially at rest. Then a strange thing happens. The moving ball (A) comes to rest

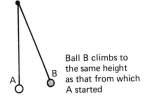

Ball A moves down to collide with B

Ball A stops and Ball B moves off

Ball B climbs to the same height as that from which A started

*Figure 1.1*

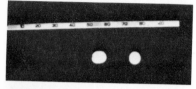

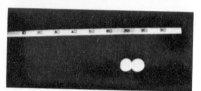

immediately on colliding with the ball which is initially at rest (B) and ball B then moves off with the same velocity as ball A had. No matter how many times you do the experiment the same thing happens.

*Figure 1.2* shows another arrangement of the same experiment. These frames from a cine film show a billiard ball colliding with another identical billiard ball that was at rest. The ball that initially was moving comes to rest after the collision while the ball that initially was at rest moves off with the same velocity the other ball had.

Why is this collision worthy of note? Why is the motion transferred from one ball to the other when they collide? Why doesn't the first ball keep some of its velocity? Why does it lose all its velocity — every time? Why does ball B, in *Figure 1.1*, rise to the same height as the ball A starts from? Why out of all the apparently possible ways in which the collision could affect the motions of the two balls does it happen in just this one way? There are a lot of questions — and the answers are not given by the conservation of momentum. The momentum of the first ball could be shared in any number of ways with the other ball — why does it always give all its momentum to the other ball?

A demonstration similar to that illustrated in *Figure 1.1* took place in the year 1666 before members of the Royal Society of London. They raised the same questions and the answers were provided by C. Huygens — not only was momentum, i.e., $mv$, conserved but there was another quantity determined by $mv^2$ which was also conserved when the collision occurred. This quantity, which we now call $\frac{1}{2}mv^2$, is known as **kinetic energy**. Momentum and kinetic energy were both conserved in the collision.

---

**Question   2**   Ball A moving with a velocity $V$ strikes ball B which is at rest in a head-on collision. After the collision we will consider ball A to have a velocity $v_A$ and ball B a velocity $v_B$. Ball A and ball B have equal masses.

(a) Write down the equation that has to be satisfied if momentum is to be conserved.

(b) The above equation does not require that $v_A = 0$ and $v_B = V$ but allows it as just one of a large range of answers. What extra condition can we apply to make certain that this answer is the only possible answer? Try squaring both sides of the equation. This should give $V^2 = v_A^2 + v_B^2 + 2v_Av_B$. What simple condition imposed on this equation will mean that either $v_A$ or $v_B$ is zero?

---

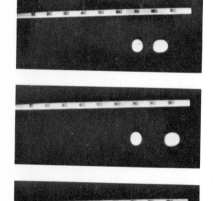

*Figure 1.2 Enlarged frames from a cine film showing a head-on collision between two identical billiard balls*

The conservation of momentum for an object of mass $m_A$ moving with a velocity $u_A$ striking an object of mass $m_B$ moving with velocity $u_B$ gives

$$m_Au_A + m_Bu_B = m_Av_A + m_Bv_B$$

where $v_A$ and $v_B$ are the velocities of A and B after the collision. The conservation of kinetic energy for the same collision gives

$$\tfrac{1}{2}m_A u_A^2 + \tfrac{1}{2}m_B u_B^2 = \tfrac{1}{2}m_A v_A^2 + \tfrac{1}{2}m_B v_B^2$$

Both equations have to be satisfied for the collisions described in
*Figure 1.1* and *Figure 1.2.*

The reason for specifying kinetic energy in this form is because it
gives a quantity which is conserved. The amount of kinetic energy
before the collision equals the amount of kinetic energy after the collision. Kinetic energy is a scalar quantity, i.e., only the size is important,
there is no significance to direction – in fact it is not possible to specify
direction. This is despite velocity being a vector quantity. Because
the kinetic energy depends on the square of the velocity the sign of the
velocity has no effect on the product. The square of a negative velocity
or a positive velocity is still a positive quantity. Momentum is a vector
quantity.

$$\text{Kinetic energy} = \tfrac{1}{2}mv^2$$

Units: mass, $m$ – kg, velocity, $v$ – m/s, kinetic energy – kg $(\text{m/s})^2$ or
kg m$^2$ s$^{-2}$ or joule (J). The unit of energy is the joule, $1 \text{ J} = 1 \text{ kg m}^2 \text{ s}^{-2}$.

---

**Questions**  **3**  Calculate the kinetic energy of a bullet of mass 2 g travelling at
400 m/s. Give the answer in joules.

**4**  A steel ball of mass 10 g moving with a velocity of 3 m/s collides
with another steel ball of mass 5 g.
If this ball is at rest when the collision occurs what will be the
velocity of each ball after the collision?

---

Is kinetic energy always conserved in collisions? If we examine a
large number of collisions the answer has to be no. When two steel
balls collide, or two billiard balls, then kinetic energy does appear to
be conserved. If, however, we consider a collision between a steel ball
and a piece of Plasticine then kinetic energy is not conserved, the total
kinetic energy after the collision is less than the initial total kinetic
energy. The steel ball and the Plasticine are very likely to stick together,
certainly the Plasticine deforms when the collision occurs. When the
collision results in one of the objects being deformed or in the two
objects sticking together then the kinetic energy after the collision is
always less than the kinetic energy before the collision. Collisions in
which kinetic energy is conserved are called **elastic collisions**, those in
which it is not conserved are termed **inelastic**. For the two steel balls,
or the two billiard balls, the collisions are such that kinetic energy is
almost completely conserved, perhaps just one or two per cent being
lost. For most purposes it is reasonable to consider such collisions to
be elastic. Truly elastic collisions occur with atomic particles (see
*Book 7: Atoms and Quanta*).

**Question    5**    A ball of mass 0.1 kg travelling with a velocity of 4 m/s collides with another ball of mass 0.2 kg. This ball was initially at rest but after the collision it becomes stuck to the other ball. What is velocity of the two balls after the collision? Is kinetic energy conserved? If not, what is the percentage loss in kinetic energy?

## Work

If an object accelerates then a force must be acting on it. The velocity of the object will be changing during the time for which the force is applied. But a change in velocity means a change in kinetic energy. Applying a force to an object and causing an acceleration causes a change in kinetic energy.

**Questions    6**    An object of mass 0.5 kg is accelerated by a force of 10 N.

(a)  What is the acceleration?

(b)  If the object had an initial velocity of 2 m/s what would the initial kinetic energy have been?

(c)  What would be the velocity after the object had been accelerated for a distance of 0.8 m?

(d)  What would be the change in kinetic energy after the object had accelerated for 0.8 m?

(e)  If the object had been accelerated for twice the distance, i.e., 1.6 m, what would the change in kinetic energy have been after the acceleration?

(f)  If the object had been accelerated for three times the distance, i.e., 2.4 m, what would the change in kinetic energy have been after the acceleration?

(g)  How would the answers to (d), (e) and (f) be changed if the force were doubled to 20 N?

(h)  How would the answers to (d), (e) and (f) be changed if the force were trebled to 30 N?

(i)  How is the change in kinetic energy related to the distance through which the object is accelerated and the force applied?

A net force $F$ produces an acceleration $a$ for an object of mass $m$ such that (*Figure 1.3*)

$$F = ma$$

The acceleration results in the velocity of the object changing from $v_0$ to $v$ in a distance $s$. These quantities are related by the equation for straight-line motion with constant acceleration such that

$$v^2 = v_0^2 + 2as$$

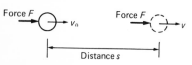

Force $F$    $v_0$        Force $F$    $v$

|← Distance $s$ →|

*Figure 1.3  A force accelerating an object*

Hence

$$v^2 - v_0^2 = 2as$$

Multiplying throughout by ½$m$ gives

$$\tfrac{1}{2}mv^2 - \tfrac{1}{2}mv_0^2 = \tfrac{1}{2}m \times 2as = mas$$

But $F = ma$, hence

$$\tfrac{1}{2}mv^2 - \tfrac{1}{2}mv_0^2 = Fs$$

The change in kinetic energy is thus determined by the net force acting on the object and the distance over which the force acts on the object. The product $Fs$ represents the energy change that occurs to the object. This product is called **work**. Work is defined as the product of the net force acting on an object and the displacement produced in the direction of the force.

$$\text{Work} = Fs$$

Units: Force, $F$ — N, distance, $s$ — m, work — N m or J. 1 J is 1 N m.

Both force and displacement are vector quantities, their magnitude and direction are both important. We are here concerned with the displacement in the direction of the force. Work is not, however, a vector quantity but a scalar quantity, as is kinetic energy.

---

**Questions**

7    A goods train which has a mass of 100 000 kg is accelerated from rest to a speed of 12 m/s in a distance of 1000 m. What is the gain in kinetic energy over that distance? What resultant force must be applied by the engine over the 1000 m? Assume that a constant force is applied.

8    A bullet hits a bank of earth and penetrates a horizontal distance of 2 m into the earth. If the bullet had a mass of 0.04 kg and a horizontal speed of 500 m/s when it hit the bank calculate

(a)  The kinetic energy of the bullet just before the impact
(b)  The average force of retardation during the bullet's passage in the earth.

---

## Power

Work describes energy in the act of being transferred from one object to another. It could be, for instance, you pulling an object. Energy is supplied by you and ends up with the object. You can calculate the amount transferred by computing the work.

There are many situations where energy is transferred from one object to another or transformed from one form to another. In order to describe the rate at which energy is transferred or transformed we use the term power.

**Power** is the rate at which energy is transferred or transformed.

Thus when the transfer is as work

Power = rate of doing work

Units: work — J, power — J/s or watt, W. 1 J/s = 1 W.

Over some interval $t$ the average power will be

$$\text{Average power} = \frac{\text{work done in time } t}{\text{time}, t}$$

As work is the product of the force and the distance through which the force acts on the object concerned,

$$\text{Average power} = \frac{\text{force} \times \text{distance covered}}{\text{time taken to cover the distance}}$$

But the distance covered divided by the time taken is the average velocity. Hence

Average power = force × velocity

---

**Questions**　**9**　A car engine develops a power of 15 kW when the car is travelling at 50 km/h.
What is the force opposing the motion?

**10**　A swimmer may use something like $10^5$ J of energy in a race lasting 30 s.
What is the swimmer's average power?

---

## Potential energy

The gravitational force acting on an object of mass $m$ is $mg$, where $g$ is the acceleration due to gravity at the point concerned. If the height of this object is changed from $h_0$ to $h$ then the height through which the object has been moved is $(h - h_0)$. This is a movement against the force $mg$. Hence work is involved.

$$\text{Work} = mg \times (h - h_0)$$
$$= mgh - mgh_0$$

But work is a measure of the change of energy of an object. Hence if we identify $mgh$ as the energy associated with the height $h$ then $mgh_0$ can be the energy associated with the height $h_0$. The energy associated with the position of an object is called **potential energy** and where the force involved is gravitational we refer to it as **gravitational potential energy**.

Gravitational potential energy = $mgh$

Units: $m$ — kg, $g$ — m/s$^2$, $h$ — m, potential energy — kg m$^2$ s$^{-2}$ or J.

**Question**  **11**    What is the potential energy of an object of mass 3 kg relative to the floor when it is 2 m above the floor?

What happens when an object falls? Because its height is decreasing its potential energy must be decreasing. The falling object accelerates, i.e., its velocity increases. This means that the kinetic energy must be increasing. When the object loses potential energy it gains kinetic energy.

**Question**  **12**    (a) What is the potential energy relative to the floor of an object of mass 2 kg at a height of 0.5 m above the floor?
(b) The object falls with an acceleration of 9.8 m/s², from rest, through a distance of 0.5 m.
What is the velocity of the object after such a distance?
(c) What is the kinetic energy gained by the object after it has fallen through 0.5 m? How does this compare with the potential energy lost as a result of the fall?

The potential energy of an object of mass $m$ at a height $h$ above some datum line is $mgh$. When the object falls from this height, initially being at rest, it accelerates with an acceleration $g$ and reaches, after the fall through the distance $h$, a velocity $v$.

The equation for motion with a constant acceleration is

$$v^2 = u^2 + 2as$$

Hence, in this particular case

$$v^2 = 2gh \quad \text{and so} \quad gh = \tfrac{1}{2}v^2$$

The potential energy lost is $mgh$. But

$$mgh = m \times \tfrac{1}{2}v^2$$

Thus the potential energy lost equals the kinetic energy gained.

**Questions**  **13**    Draw graphs showing how the potential energy and the kinetic energy of a freely falling object change with the distance fallen from the rest position. How does the total energy vary with the distance fallen? The total energy is the sum of the potential and kinetic energies.

**14**    *Figure 1.4* shows a simple pendulum. When the pendulum is pulled to one side the bob of the pendulum is lifted through a vertical height. When the pendulum swings the bob first loses that height and then regains it.

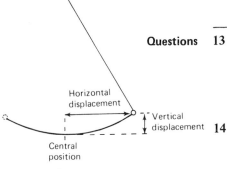

Horizontal displacement

Vertical displacement

Central position

*Figure 1.4*

**Questions
continued**

Draw graphs showing how

(a) the potential energy
(b) the kinetic energy
(c) the total energy

vary with the horizontal displacement of the pendulum bob
from its central position.

---

Distance through which mass is pulled $(x - x_0)$

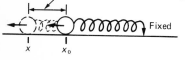

Distance

*Figure 1.5*

There are other instances where an object can have energy by virtue
of its position; instances which do not depend on gravitational forces.
Suppose you pulled an object which was tethered to a horizontal spring,
as in *Figure 1.5*. To move the object through some distance you have
to pull it against a force, a force which wants to pull the object back
to its initial position. This is a similar situation to lifting an object
through a vertical distance against the gravitational force which wants
to pull the object back to its initial position. In that situation we considered that the gravitational force remained constant over the distance
through which the object was lifted. We cannot make this assumption
for the elastic force of the spring, the force increases as the spring is
extended. The greater the extension of the spring the greater the force
acting on the object. The force is proportional to the extension. *Figure
1.6* shows a graph of the force plotted against the extension, the graph
being described by the relationship $F = kx$.

When the spring extends from extension $x_0$ to $x$ the force changes
uniformly from $F_0$ to $F$. The average force during this extension is

$$\text{Average force} = \frac{F_0 + F}{2}$$

However, $F_0 = kx_0$ and $F = kx$, hence

$$\text{Average force} = \frac{kx_0 + kx}{2}$$

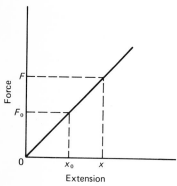

*Figure 1.6*

The work involved as the force moves the object at the end of the
spring from $x_0$ to $x$ is therefore

$$\begin{aligned}
\text{Work} &= \left(\frac{kx_0 + kx}{2}\right) \times (x - x_0) \\
&= \tfrac{1}{2}k(x + x_0)(x - x_0) \\
&= \tfrac{1}{2}k(x^2 - x_0^2) \\
&= \tfrac{1}{2}kx^2 - \tfrac{1}{2}kx_0^2
\end{aligned}$$

Work represents the input of energy to the system when the mass is
pulled from a distance $x_0$ to $x$. There would thus seem to be an energy
of $\tfrac{1}{2}kx^2$ associated with the position of the object. This energy is often
known as **elastic potential energy.**

Elastic potential energy $= \tfrac{1}{2}kx^2$

Units: $k$ – N/m, $x$ – m, elastic potential energy – N m or J.

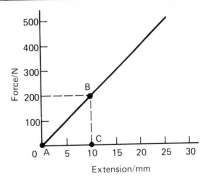

*Figure 1.7*

**Questions** 15    *Figure 1.7* shows a graph of the force applied to stretch a spring and the extension produced by that force.

(a) What is the force when there is zero extension?
(b) What is the force when there is an extension of 10 mm?
(c) What is the average force during the extension from zero to 10 mm?
(d) What is the energy needed to increase the extension from zero to 10 mm?
(e) Calculate the area of the triangle ABC. How does the area compare with the answer given in part (d)?
(f) *Figure 1.8* shows a force/extension graph for another spring where the extension is not always proportional to the force. What is the work involved in stretching the spring from 0 to 10 mm? What is the work involved in stretching the spring from 10 to 20 mm? What is the work involved in stretching the spring from 0 to 20 mm? How do the answers obtained by calculating the average forces for each part of the extension compare with that obtained by estimating or calculating the area under the graph?

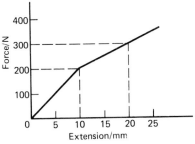

*Figure 1.8 The area under a force/extension graph is the energy stored in the material*

16    A spring is held compressed by a force of 200 N. To reach that state the spring was compressed through a distance of 50 mm. What is the potential energy stored in the spring? Assume that the spring obeys Hooke's law, i.e., compression is proportional to the applied force.

**17**   The spring in an air rifle is normally compressed by 50 mm. When the rifle is cocked the spring is compressed by a further 50 mm to a total compression of 100 mm.

If the spring obeys Hooke's law and has a force constant of 2000 N/m, i.e., this is the value of $k$ in $F = kx$, what is the potential energy added to the spring by cocking the rifle? If all this potential energy is converted to kinetic energy for the shot, mass 0.4 g, what will be its velocity?

From the previous discussion it is apparent that potential energy can be stored in a coiled spring. But this is only a special case of the behaviour of any solid when under stress. If you stretch a rubber band and it suddenly snaps, the broken end can whip back and give you a nasty sting. When the rubber is stretched, potential energy is being stored up in the rubber and this is released when the tension ceases and appears as kinetic energy. A steel hawser under tension can have a considerable amount of energy stored in it, as potential energy, and if the end of the hawser is suddenly released the free end can whip back and cause considerable damage. If you do an experiment involving the stretching of a steel wire you must take care not to be near the wire if it suddenly breaks or the load slips off.

The energy stored in a stretched strip of material is equal to the average force causing the material to stretch multiplied by the distance through which the material has been extended as a result of the force, assuming Hooke's law to be obeyed. Thus if the force is increased from zero to $F$, the average force is $\frac{1}{2}F$.

$$\text{Work } = \text{ average force} \times \text{extension } = \frac{1}{2}\text{ force} \times \text{extension}$$

or

$$\text{Energy stored } = \frac{1}{2}\text{ force} \times \text{extension}$$

Hence

$$\frac{\text{Energy stored}}{\text{Volume of material}} = \frac{\frac{1}{2}\text{ force} \times \text{extension}}{\text{volume of material}}$$

$$\text{Energy stored per unit volume } = \frac{\frac{1}{2}\text{ force} \times \text{extension}}{\text{area} \times \text{length}}$$

But

$$\frac{\text{Force}}{\text{Area}} = \text{stress} \quad \text{and} \quad \frac{\text{extension}}{\text{length}} = \text{strain}$$

Hence

$$\text{Energy stored per unit volume } = \frac{1}{2}\text{ stress} \times \text{strain}$$

This stored energy can be called elastic potential energy but more often is called **strain energy** (*Figure 1.9*).

The stress and strain over the region for which Hooke's law applies are related by

$$\frac{\text{Stress}}{\text{Strain}} = \text{Young's modulus}$$

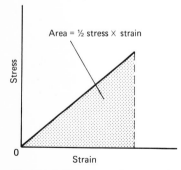

Area = ½ stress × strain

Stress

0          Strain

*Figure 1.9 The area under a stress/strain graph is the energy stored per unit volume*

Steel has, typically, a modulus of elasticity of $220 \times 10^9$ N/m² and may be subject within its elastic region to strains up to 0.001 (*see* Chapter 1, *Book 2: Materials*). The strain energy per unit volume for this strain is thus $\frac{1}{2} \times 0.001 \times 220 \times 10^9 \times 0.001$ or 110 000 J/m³.

**Questions**

**18** Rubber has a Young's modulus of $7 \times 10^6$ N/m² and is used up to a strain of 0.5.
Assume that the stress is proportional to the strain over this region and calculate the energy stored per unit volume for this strain. What is the energy stored in a piece of this rubber of length 100 mm and cross-section 5 mm by 5 mm when subject to this strain?

**19** What would be the energy stored in a piece of steel of the same size as the piece of rubber in question **18**? Take the data to be as given prior to this question, the strain thus being 0.001.

**20** Car bumpers are often made of steel, though some cars have a rubber bumper wrapped round the front of the car. When a car collides with some object the kinetic energy of the car is transformed to strain energy within the bumper, assuming no deformation of the car or bumper takes place.
Which material, in view of your answers to questions 18 and 19 should be better able to cushion the effect of a collision by absorbing the kinetic energy?

**21** What characteristics would you expect of a material used as packing material in a box which is to carry a delicate piece of scientific equipment?

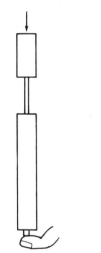

*Figure 1.10 Squashing air in a bicycle pump*

If you take a bicycle pump, put your finger over the valve connection and then depress the pump handle, there is opposition because of the increase in pressure, and the decrease in volume, of the air trapped in the pump (*Figure 1.10*). If after 'squashing' the air in the pump you let go of the handle then the compressed air pushes the handle back out again. It is just like applying a force to a piece of rubber to squash it — when you release the rubber by removing the force the rubber springs back to its original shape (provided it is not stressed beyond the elastic limit). The rubber has elastic potential energy when squashed, work is done in squashing the rubber. It thus seems reasonable to talk of the compressed gas having potential energy and consider the work involved in compressing it.

*Figure 1.11* shows a gas being compressed in a cylinder by a piston, as in the bicycle pump. If a force $F$ is applied to the piston and it moves a small distance $\Delta L$ then the work done is

$$\text{Work done} = F \times \Delta L$$

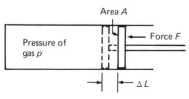

*Figure 1.11*

But this force $F$ acts over a piston area $A$ thus the pressure $p$ applied to the gas is

$$p = F/A$$

Hence

$$\text{Work done} = pA \times \Delta L$$

But $A \times \Delta L$ is the change in volume $\Delta V$. Hence

$$\text{Work done} = p \times \Delta V$$

The reason for considering the change in volume, and the distance moved by the piston, to be small is that the force needed to move the piston and compress the gas will not be constant as the volume is progressively reduced.

The work is the energy transferred to the gas. This is potential energy stored in the gas as a result of its volume having been reduced by $\Delta V$ at a pressure $p$.

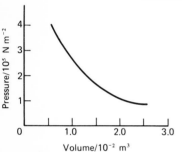

Figure 1.12

**Question 22**   *Figure 1.12* shows how the pressure of a enclosed volume of gas depends on its volume when no temperature change occurs.

(a)  What energy is transferred to the gas when the volume is decreased from $2.5 \times 10^{-2}$ m³ to $2.0 \times 10^{-2}$ m³? Assume that the pressure does not change significantly during this volume change and is considered to be $1 \times 10^5$ N/m².

(b)  Your answer to (a) is the area under the graph between $2.5 \times 10^{-2}$ m³ and $2.0 \times 10^{-2}$ m³. Explain why this is the case.

(c)  What energy is transferred to the gas when the volume is reduced from $2.5 \times 10^{-2}$ m³ to $0.5 \times 10^{-2}$ m³? The pressure cannot be assumed to remain constant during the change.

## Energy of rotation

When a force is used to rotate an object (*Figure 1.13*) work is involved in that a force is being used to move an object. Thus when the tangential force $F$ acts at a point A and causes it to move through an arc AB then the work involved is

$$\text{Work} = F \times \text{distance round arc AB}$$

But

$$\text{Arc AB} = r\theta$$

where $\theta$ is the angle swept out by the movement through the arc AB (see *Book 1: Motion and Force* for more information on rotation). Thus, the work is given by

$$\text{Work} = F \times r\theta$$

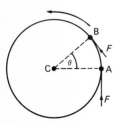

Figure 1.13  *Force causing an object to rotate about C*

But the torque exerted by the force is the product of the force and the radius of the rotation caused by the force.

Torque, $T = Fr$

Hence

Work $= T\theta$

Units: $T$ – N m, $\theta$ – radians, work – J.

If the object rotates through $f$ revolutions in one second then in one second the angle turned through will be $2\pi f$, there being $2\pi$ radians to one complete revolution. Hence, the work done in one second on an object rotating at a frequency of $f$ revolutions per second is

Work done per second $= T \times 2\pi f$

This work done per second is the rate at which energy is transferred and so is the power.

Power $= T \times 2\pi f$

Units: power – J/s or W, $T$ – N m, $f$ – s$^{-1}$.

---

**Question** **23** Calculate the power input needed to keep a wheel rotating at 5 revolutions per second when acted on by a constant torque of 40 N m.

---

An energy input is needed to make a wheel, or indeed any object, rotate. The energy input, when there is a constant torque, $T$, is the work $T\theta$. When there is an energy input to an object which is moving in a straight line the object gains in velocity and we talk of the object having acquired energy. Thus, if the energy input causes the velocity to increase we say that the work has resulted in the object gaining kinetic energy. Similarly when there is an energy input which results in an object rotating then we talk of the object gaining **rotational kinetic energy**.

When a constant torque acts on an object then there is a constant angular acceleration, i.e., the angular velocity increases at a constant rate.

Torque, $T = I\alpha$

where $I$ is the moment of inertia and $\alpha$ the angular acceleration. Thus, the work input which results in this change in angular velocity is

Work $= I\alpha \times \theta$

The angular acceleration $\alpha$ when the angular velocity changes at a constant rate from $\omega_0$ to $\omega_1$ in a time $t$ is

$$\alpha = \frac{\omega_1 - \omega_0}{t}$$

The angle $\theta$ rotated through in a time $t$ is the product of the average angular velocity and the time $t$.

$$\text{Average angular velocity} = \frac{\omega_1 + \omega_0}{2}$$

Hence the work is

$$\text{Work} = I \times \frac{\omega_1 - \omega_0}{t} \times \frac{\omega_1 + \omega_0}{2} \times t$$

$$\text{Work} = \tfrac{1}{2} I \omega_1^2 - \tfrac{1}{2} I \omega_0^2$$

Work is a measure of the change in energy of a body. It is thus reasonable to consider that an object rotating with an angular velocity $\omega$ has a rotational kinetic energy given by

$$\text{Rotational kinetic energy} = \tfrac{1}{2} I \omega^2$$

Units: rotational kinetic energy $- \text{J}, I - \text{kg m}^2, \omega - \text{s}^{-1}$.

This equation compares with the kinetic energy for translational motion of $\tfrac{1}{2} m v^2$; the mass term in translational motion has the moment of inertia as its equivalent in rotational motion and the velocity term in translational motion has angular velocity as its equivalent in rotational motion.

**Question  24**   A wheel has a moment of inertia of 50 kg m$^2$ and is rotating at 2 revolutions per second.
What is the rotational kinetic energy of the wheel? Note: $\omega = 2\pi f$ (*see Book 1: Motion and Force*).

A large rotating object is an energy store of, possibly, quite a large capacity. Many machines use flywheels as energy stores. A flywheel could quite simply be a disc with its axis coincident with the axis of a drive shaft from some motor. If the energy input to the motor increases the rotational speed of its drive shaft will increase. If, however, there is a flywheel, energy is 'absorbed' by the flywheel with very little increase in speed. This is because the rotational kinetic energy of the flywheel is dependent on the square of the angular velocity and the moment of inertia term can be made very large. Similarly if the energy input to the motor decreases energy can be taken from the flywheel without much of a drop in speed. The flywheel enables the speed of the motor to remain reasonably constant despite changes in the energy input.

**Question  25**   A flywheel has a moment of inertia of 300 kg m$^2$ and is rotating at 3 revolutions per second.
What is the energy stored in the flywheel? By what factor has the energy input to change to give a 5% change in angular velocity?

## Heat energy

On a cold day you might rub your hands together to make them warm.
Where does this rise in temperature come from? In the early part of
the 19th century the answer would probably have been — you were
squeezing 'caloric' out of your hands by the rubbing. This theory was
known as the **caloric theory**. Caloric was regarded as substance. Now
we would notice the connection between the motion and the rise in
temperature and talk of the kinetic energy of the motion of the hands
being transformed into heat energy.

The crucial series of experiments that, probably, led to heat being
recognised as a form of energy were those of James Joule in the middle
of the 19th century. *Figure 1.14* shows the basic form of apparatus he
used for some of the experiments. It consists of a paddle wheel in
water. The paddle wheel is rotated as a result of a mass falling. The
falling mass loses potential energy and produces a rise in the temperature of the water in the container as a result of the paddle wheel
'rubbing' against the water. Joule found that equal temperature rises
were produced per unit mass of water for equal expenditures of potential energy by the falling mass. There was, therefore, a direct equivalence
between potential energy and heat — heat would, therefore, appear to
be a form of energy.

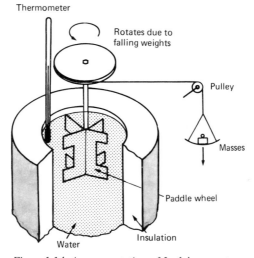

*Figure 1.14  A representation of Joule's apparatus*

Joule suggested a natural phenomenon where the conversion of
potential energy to kinetic energy and hence heat could be observed
and measured (*Philosophical Magazine*, 1845).

> 'Any of your readers who are so fortunate as to reside amid
> the romantic scenery of Wales or Scotland, could, I doubt not,
> confirm my experiments by trying the temperature of the water
> at the top and at the bottom of a cascade [waterfall] . If my views
> are correct, a fall of 817 feet [about 249 m] will of course gener-
> ate one degree of heat [about 0.56 °C] ; and the temperature of
> the river Niagara [the Niagara falls] will be raised about one fifth
> of a degree by its fall of 160 feet [about 49 m] .'

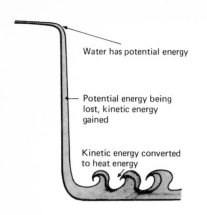

Water has potential energy

Potential energy being
lost, kinetic energy
gained

Kinetic energy converted
to heat energy

*Figure 1.15*

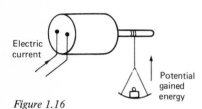

Electric
current

Potential
gained
energy

*Figure 1.16*

*Figure 1.15* shows the sequence as envisaged by Joule: potential energy to kinetic energy to heat. William Thomson, later known as Lord Kelvin, when walking near Chamonix in 1847 reported

'. . . and whom should I meet walking up but Joule, with a long thermometer in his hand, and a carriage with a lady in it not far off. He told me that he had been married since we parted in Oxford [two weeks earlier] and he was going to try for elevation of temperature in waterfalls.'

That was on his honeymoon. The effect does exist — the temperature of the water at the base of a waterfall is higher than the temperature at the top and the effect is due to the transformation of potential energy to kinetic energy and hence heat.

## Electrical energy

When a current passes through an electric motor and the motor shaft rotates energy is involved. The motor might, for instance, be used to lift a load (*Figure 1.16*). The load is given potential energy. When a current passes through an immersion heater the heater becomes warm. Energy is involved. It would appear that electricity supplies energy.

**Question    26**    A hydroelectric station (see the photograph at the beginning of this chapter) has an input of energy in the form of the potential energy of water and an output of energy as electricity. A particular hydroelectric station has a water reservoir 720 m above its turbines. The water from this reservoir falls through tunnels, losing its potential energy and gaining kinetic energy. When the water hits the blades of a turbine the kinetic energy of the falling water is transformed into rotational kinetic energy. This in turn is transformed into electricity by a generator. The hydroelectric station has a number of turbines, and each turbine produces a power of 56 MW.

(a) Assuming all the kinetic energy of the falling water is converted into electrical energy, how much kinetic energy does the water 'lose' to a turbine per second?
(b) What mass of water hits the blades of a turbine per second?
(c) With what velocity does the water hit a turbine blade?

## Forms of energy

There are many situations in which we can recognize that energy is involved. For instance: kinetic energy of moving bodies, gravitational potential energy, elastic or strain energy, heat, electrical energy, chemical energy, nuclear energy. You can no doubt add more to the list. We recognize energy as being involved because we have inputs and outputs to a system and, if we consider these in terms of energy, we can 'balance' the energy account 'books' in terms of a common 'currency' which is involved in all the changes — that 'currency' is energy. There are many monetary currencies in the world, pounds, dollars, francs, etc. We recognize them all as money because we can change from one to another — for international purposes we can talk of all of them in terms of a common currency, the dollar, and indeed most international negotiations are carried out in terms of the dollar though the currency that may actually be used for payment is different.

---

**Questions**   27   Consider the following systems — what forms do the input and output energies take?

(a) A car gearbox.
(b) A lever.
(c) A loudspeaker.
(d) A battery.
(e) A coal-fired boiler.
(f) An electric fire.
(g) A microphone.
(h) An electric lamp.
(i) An electric motor.

28   In your own words explain what is meant by 'energy is always conserved'.

---

### The first law of thermodynamics

If a system is completely isolated in such a way that no energy enters or leaves it then the total energy of that system must remain constant.

Initial energy of system  =  Final energy of system

The term **system** is here used to describe a collection of matter within some prescribed boundaries — we could, for instance, be considering the system to be a piston enclosing gas within a cylinder.

If a system is not isolated and energy flows into and/or out of it then

Initial energy of system + energy entering system  =
final energy of system + energy leaving the system

The energy transfer into the system or out of the system takes place as work or heat. If we consider there to be an energy inflow, as heat, of $Q$ and an energy outflow, as work, of $W$ then

$$U_i + Q = U_f + W$$

where $U_i$ is the initial energy of the system and $U_f$ the final energy. Rearranging this equation gives

$$U_f - U_i = Q - W$$

The change in energy of the system is $(U_f - U_i)$ and can be represented as $\Delta U$.

$$\Delta U = Q - W$$

The energy contained within a system is called the **internal energy** of the system. Thus $U$ represents the internal energy and $\Delta U$ the change in internal energy.

The above equation describes what is known as the **first law of thermodynamics.** The change in internal energy of a system is equal to the heat absorbed minus the external work done by the system.

In the above argument both heat and work are considered to be just the means of communicating energy. The result of the communication may be a change in the internal energy of the system. Heat and work are ways of changing internal energy not quantities that a system can be said to 'contain'. It is like conversation between two people. The conversation is a means of communication between the two but we would not state that the people at the end of their interaction 'contained' conversation.

In 1843 Joule tried an experiment in which he allowed a gas in one enclosure to expand into a vacuum in another enclosure (*Figure 1.17*). The entire system was lagged. When the expansion occurred no temperature change was detected. From the first law of thermodynamics we must conclude that as the expanding gas was doing no work and no heat was being exchanged with any other system, the internal energy of the gas must remain unchanged. No work is done by the gas because it is 'pushing' against nothing when it expands. No change in internal energy and no change in temperature, but a change in volume and a change in pressure, leads us to infer that the internal energy of the gas is related only to the temperature.

Experiments later than that of Joule have shown that for real gases there is a small temperature change when gases expand freely into a vacuum. The condition of no temperature change is considered to apply only to an ideal gas. Thus for an ideal gas the temperature is a measure of the internal energy.

The internal energy of liquids and solids does not depend only on the temperature. The reason for real gases, liquids and solids having internal energies dependent on factors additional to temperature is that there is energy locked up in such materials in the form of potential energy due to the bonding between the constituent particles, i.e., inter-particle forces. We can think of the ideal gas as being a collection of particles with no inter-particle forces.

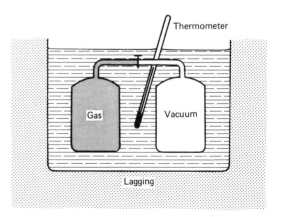

*Figure 1.17*

**Suggestions for answers**  1    Take as an example the reaction

$$C + O_2 \rightarrow CO_2$$

One atom of carbon combines with one molecule, two atoms, of oxygen to give one molecule of carbon dioxide. This molecule contains one atom of carbon for every two atoms of oxygen. In the reaction the total number of carbon and oxygen atoms after the reaction is the same as the number prior to the reaction.

2    (a) $V = v_A + v_B$
(b) $v_A$ or $v_B$ will be zero if $V^2 = v_A^2 + v_B^2$

3    $\frac{1}{2} \times 2 \times 10^{-3} \times 400^2 = 160$ J

4    $\frac{1}{2} \times 0.01 \times 3^2 = \frac{1}{2} \times 0.01 \times v_1^2 + \frac{1}{2} \times 0.005 \times v_2^2$ and so $9 = v_1^2 + 0.5v_2^2$. $0.01 \times 3 = 0.01v_1 + 0.005v_2$ by the conservation of momentum. Thus $3 = v_1 + 0.5v_2$ and so $9 = (3 - 0.5v_2)^2 + 0.5v_2^2$. Hence $v_2 = 4$ m/s and $v_1 = 1$ m/s.

5    $0.1 \times 4 = 0.3v$, hence $v = \frac{4}{3}$ m/s. Initial kinetic energy is 0.8 J. Final kinetic energy is about 0.27 J. Kinetic energy is therefore not conserved, and the percentage loss is about 66%.

6    (a) 20 m/s$^2$
(b) 1 J
(c) $v^2 = u^2 + 2as = 36$, hence $v = 6$ m/s.
(d) $9 - 1 = 8$ J
(e) 16 J
(f) 24 J
(g) Change in kinetic energy doubled
(h) Change in kinetic energy trebled
(i) Change in kinetic energy is proportional to the product of the force and the distance.

7    Gain in kinetic energy $= 72 \times 10^5$ J. Thus as $\frac{1}{2}mv^2 = Fs$, the force is $72 \times 10^2$ J.

8    (a) $5 \times 10^3$ J
(b) $2.5 \times 10^3$ N

9    $15 \times 10^3 = F \times 50 \times 10^3/3600$, hence $F = 1080$ N

10    $10^5/30$ W

11    $mgh = 3 \times 9.8 \times 2 = 58.8$ J

12    (a) 9.8 J
(b) $v^2 = u^2 + 2as$, hence $v = (9.8)^{\frac{1}{2}}$ m/s
(c) 9.8 J. Kinetic energy gained $=$ potential energy lost

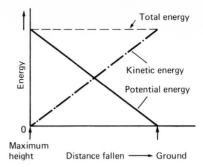

*Figure 1.18*

**13**   See *Figure 1.18*. The potential energy is directly proportional to the distance above the ground, being *mgh*. The kinetic energy is proportional to $v^2$, being $\frac{1}{2}mv^2$. However as $v^2 = 2gh$ the kinetic energy is directly proportional to the distance fallen. As the potential energy decreases the kinetic energy increases, the total of the two remaining constant.

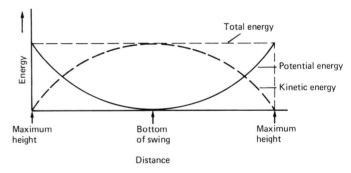

*Figure 1.19*

**14**   See *Figure 1.19*. The total of the kinetic energy and the potential energy at any distance is a constant. The curvature of the potential and kinetic energy graphs is the same as that of the pendulum path.

**15**   (a)  0
     (b)  200 N
     (c)  100 N
     (d)  $100 \times 0.010 = 1$ J
     (e)  Area $= \frac{1}{2} \times 200 \times 0.010 = 1$ J. Area under the graph is the same as change in energy in (d)
     (f)  1.0 J, 2.5 J, 3.5 J. Work $=$ area under the graph

**16**   $\frac{1}{2} \times 200 \times 0.050 = 5$ J

**17**   $150 \times 0.050 = 7.5$ J, about 194 m/s

**18**   $\frac{1}{2}$ stress $\times$ strain $= \frac{1}{2}E \times$ strain$^2 = 8.75 \times 10^6$ J/m$^3$. About 21.9 J

**19**    About 0.275 J

**20**    The rubber

**21**    Ability to absorb a high amount of energy, i.e., large energy per unit volume.

**22**    (a)  $1 \times 10^5 \times 0.5 \times 10^{-2} = 0.5 \times 10^3$ J
(b) Area = work. The sum of all the $p\Delta V$ terms is the area under the graph.
(c) Approximately $3.2 \times 10^3$ J

**23**    Power = $40 \times 2\pi \times 5$ or about 1260 W

**24**    Rotational kinetic energy = $\frac{1}{2} \times 50 \times (2\pi \times 2)^2$, about 3.9 kW

**25**    Rotational kinetic energy = $\frac{1}{2} \times 300 \times (2\pi \times 3)^2$, about 53 kW. As the energy is related to the square of the angular velocity a 5% change in angular velocity will require a 10% change in energy.

**26**    (a) 56 MJ
(b) $mgh$ = 56 MJ, hence $m$ is about $7.9 \times 10^3$ kg
(c) $mgh = \frac{1}{2}mv^2$, hence $v$ is about 119 m/s

**27**    (a) Both mechanical, (b) both mechanical, (c) electrical to sound, (d) chemical to electrical, (e) chemical to heat, (f) electrical to heat, (g) sound to electrical, (h) electrical to light, (i) electrical to mechanical.

**28**    See Appendix 1.1 for some ideas

---

**Further problems**    No suggestions for answers are given for these problems.

**29**    Explain the term energy to someone who has not studied science.

**30**    Explain the terms potential energy and kinetic energy.

**31**    A steel ball of mass 2 g moving with a velocity of 4 m/s collides with another steel ball of mass 4 g.
If this other ball is at rest when the collision occurs what will be the velocity of each ball after the collision? Assume that kinetic energy is conserved.

**32**    Consider two objects with masses $m_1$ and $m_2$, velocities $v_1$ and $v_2$ respectively.

(a) What is the total kinetic energy?
(b) The two objects collide. After the collision the two objects have velocities $w_1$ and $w_2$. What is the total kinetic energy after the collision?
(c) If the collision was perfectly elastic how would the total kinetic energy before the collision be related to the total after the collision?
(d) If the change is not perfectly elastic, let $E$ be the change in energy, how is your answer to (c) modified?

(e)  Now view the collision from another frame of reference, one moving with a constant velocity, $V$. Then from this point of view velocity $v_1$ becomes $v_1 - V$, velocity $v_2$ becomes $v_2 - V$, velocity $w_1$ becomes $w_1 - V$, velocity $w_2$ becomes $w_2 - V$. The change in energy $E$ remains the same.

What is the equation given by applying the conservation of energy?

(f)  The answers to (d) and (e) must be identical if we assume that the law of conservation of energy does not change when we observe from a frame of reference moving with a different uniform velocity.

What is the condition for this to be true? Have you met this condition before — what is it?

**33**  An object of mass 5 kg moving due east at 2 m/s strikes an object of mass 3 kg moving due west at 2 m/s.

If the collision is elastic what are the velocities of the two objects after the collision?

**34**  Prove that when an object makes a completely elastic collision with another object of the same mass but initially at rest one of the objects must be at rest after the collision.

**35**  Which of the following collisions are elastic and which inelastic?

(a)  An object of mass 2 kg moving at 4 m/s collides with an object of mass 2 kg at rest. After the collision both objects proceed in the same direction as the first object was moving before the collision but with the first object having a velocity of 1 m/s and the other a velocity of 3 m/s.

(b)  An object of mass 2 kg moving at 4 m/s collides with an object of mass 2 kg at rest. After the collision the first object is at rest and the other object moves with a velocity of 4 m/s in the same direction as the initial velocity.

(c)  An object of mass 2 kg moving at 4 m/s collides with an object of mass 2 kg at rest. After the collision the first object bounces back with a velocity of 1 m/s and the other moves with a velocity of 5 m/s in the same direction as the initial velocity.

**36**  A 100 kg meteorite moving with a velocity of 80 m/s relative to the earth hits the earth. The mass of the earth is $6 \times 10^{25}$ kg.

(a)  What is the momentum of the meteorite relative to the earth?
(b)  What is the momentum of the combined meteorite–earth system after the collision?
(c)  What is the velocity gained by the earth as a result of the collision?
(d)  What was the initial kinetic energy of the meteorite?
(e)  What is the kinetic energy of the earth–meteorite system after the collision (do not assume the collision to be elastic)?
(f)  What fraction of the initial kinetic energy is used to warm up the earth and the meteorite?
(g)  You should now be able to answer the question of why many meteorites burn up on hitting the earth's surface.

37 It has been suggested that work is analogous to the spending of money. Consider the analogy and what would represent energy in this analogy.

38 When an object is pulled along the floor with a constant velocity by means of a rope the tension in the rope is 400 N.
What is the work when the object is pulled a distance of 5 m?

39 A bullet of mass 2 g moving with a velocity of 300 m/s hits a tree. It comes to rest after penetrating to a depth of 50 mm.
What was the average force of retardation during the passage of the bullet within the tree?

40 What is the change in the gravitational potential energy of a stone of mass 50 kg when it is lifted through a vertical height of 3 m?

41 A jet airliner has a mass of $70 \times 10^3$ kg, when fully loaded.
What energy is needed to lift the airliner to its cruising altitude of 10 000 m? Assume that the acceleration due to gravity is constant over that distance and has a value of 10 m/s$^2$.

42 A spring placed vertically on a bench is compressed by 10 mm and while in the compressed state has a small ball of mass 0.01 kg placed on top of it. When the spring is released the ball shoots up in the air, to a height of 200 mm above the top of the spring.
What was the average force used to compress the spring?

43 What work is necessary to compress a spring having a force constant of 500 N/m from zero compression to 50 mm compression?

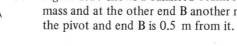

*Figure 1.20*

44 *Figure 1.20* shows a balanced beam. At end A there is a 4 kg mass and at the other end B another mass. End A is 1.0 m from the pivot and end B is 0.5 m from it.

(a) If end A moves upwards by 0.2 m, by how much is end B depressed? Assume that the beam remains on the pivot and that the pivot does not move. Also assume that the beam does not bend.
(b) What would the mass at end B have to be if there was to be no net gain in potential energy when A moves upwards?
(c) Suppose A moved through a different distance, should the mass at B have the same value as in (b) for there to be no net gain in potential energy?
(d) When a system does not gain in potential energy when displaced then it is said to be in equilibrium. Write out an algebraic equation for the potential energy sum when the beam is in equilibrium, i.e., balanced.
(e) How are the vertical distances through which A and B are displaced related to the distances of A and B from the pivot?
(f) Use your answers to (d) and (e) to arrive at an equation relating the forces applied at A and B and the distances of the points of application of these forces from the pivot.

(g) The product of the force and the perpendicular distance from the line of action of the force to the pivot is called the moment of the force. When a beam is in equilibrium there is no net moment. Does this check with your answer to (f)?

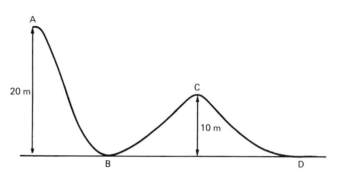

*Figure 1.21*

45    *Figure 1.21* shows the side view of a roller coaster track. The car of mass 300 kg, including the passengers, starts with zero velocity at A.

(a) What will be the speed of the car at B and at C?
(b) At the end of the track the car runs against brakes at D. What energy will have to be 'lost' by the car for it to come to a stop at D?

46    As part of an investigation a student devises a machine for throwing a ball vertically upwards with controlled and measured velocities. He intends to determine how the loss of energy due to air resistance depends on the initial velocity of the ball. The only measurement that he intends to make other than that of the initial velocity and mass of the ball is the maximum height reached by the ball.
Explain how he would calculate the energy loss.

47    The rubber of the typical rubber band used to keep papers together has an average Young's modulus of about $7 \times 10^6$ N/m². What is the energy stored in the rubber band under typical extensions? Find the other data you need for this question.

48    Explain how you could set up an investigation to determine the power generated by a student riding a bicycle.

49    A motor boat when going flat out has a constant speed of 8 km/h. When running at this speed the engine delivers energy at 3 kW.
What is the average force resisting the motion of the boat at this speed?

50    The hammer of a pile driver has a mass of 500 kg and has to be lifted through a vertical distance of 2 m in 4 s before it is dropped onto the pile.
What is the power required of the engine lifting the hammer?

51    A man climbs the stairs to the fourth floor of a building in 28 s. If this is a vertical height of 12 m what was his power?

52    What is the power needed for an engine which has to pump 0.6 m$^3$ of water per minute from a well 8 m deep and eject the water with a speed of 5 m/s?

53    State the energy transformations involved in the hydroelectric station shown in the opening photograph to this chapter.

54    What is the power needed to give a rotation of 4 revolutions per second when there is a torque of 60 N m?

55    An electric motor has a drive shaft which rotates at 20 revolutions per second and gives an output of 3 kW.
What is the output torque of the motor?

56    Calculate the gain in rotational kinetic energy of a flywheel of moment of inertia 20 kg m$^2$ when it is accelerated from 3 to 5 revolutions per second.

57    What is the rotational kinetic energy of a wheel of moment of inertia 160 kg m$^2$ rotating at 160 radians per second?

## Appendix 1.1   Some views on energy conservation

The following are the ways a number of authors have considered the conservation of energy. You might like to consider what your view is.

There is a fact, or if you wish, a law, governing all natural phenomena that are known to date. There is no known exception to this law — it is exact so far as we know. The law is called conservation of energy. It states that there is a certain quantity, which we call energy, that does not change in the manifold changes which nature undergoes. That is a most abstract idea, because it is a mathematical principle; it says that there is a numerical quantity which does not change when something happens. It is not a description of a mechanism, or anything concrete; it is just a strange fact that we can calculate some number and when we finish watching nature go through her tricks and calculate the number again, it is the same. (Something like the bishop on a red square, and after a number of moves — details unknown — it is still on some red square. It is a law of this nature.)

*The Feynman Lectures on Physics*, Vol. 1, by R.P. Feynman, R.B. Leighton and M. Sands.

Inside an isolated system, or universe, the sum of the energies of the parts remains constant irrespective of any changes that occur there. This fact is frequently expressed by saying that energy is conserved.

(Most people seem to believe this firmly; mathematicians because they believe it is a fact of observation; observers because they believe it is a theorem of mathematics; philosophers because they believe it is aesthetically satisfying, or because they believe no inference based upon it has ever been proven false, or because they believe new forms of energy can always be invented to make it true. A few neither believe nor disbelieve it; these people maintain that the First Law is a procedure for bookkeeping energy changes, and about bookkeeping procedures it should be asked, not are they true or false, but are they useful.)

*The Second Law* by H.A. Bent.


## Vicissitudes of conservation

The complete meaning of the conservation of energy was not recognized until the middle of the nineteenth century. This is rather strang considering that Newton had recognized the conservation of momentum two centuries earlier. But in the early days of the modern era of science there was great confusion over many matters which every beginning physics student is expected to understand clearly these days. Gottfried Wilhelm von Leibniz and René Descartes (together with their followers) fought for many years over the question of 'conservation of force'. Descartes chose to measure force by the product *mv*, which he called the quantity of motion, and he asserted that the total quantity of motion in the universe must remain constant. Leibniz countered this view (in 1686) by saying that a force must be measured by the *vis viva* (energy in motion) produced in a body when the force acted upon it through a certain distance.

It was not until 1743 that Jean d'Alembert pointed out that everybody was just arguing about words — a force could be measured either by the momentum it gave a body or by the energy. In other words, there were actually *two* conservation laws involved during all the quarrelling, and both sides were right.

Actually, the general idea that the universe contains a constant amount of motion, or force, or some other dimly understood form of energy goes back a long way. This undoubtedly is connected with the hundreds of years of futile attempts to create a perpetual-motion machine — a machine that would keep running and do work without the benefit of some means of propulsion. The idea of the overbalancing wheel goes back to the thirteenth century and was foisted on gullible believers as late as the nineteenth. (If it has succeeded in fooling anybody in the twentieth century, it has not come to my attention.)

The failure of all attempts to make a successful perpetual-motion machine must have convinced many that it was impossible to get mechanical energy for nothing. Newton, Leibniz, and some of their contemporaries based their work on this conviction. However, it was not so clear to those early thinkers that *all kinds* of energy were conserved. It was possible for a person to think that a mechanical perpetual-motion machine (like the overbalanced wheel) would not work, while at the same time believing that an electric, chemical, or heat device might somehow create energy.

We must remember that it was not then realized that energy could be changed from one form to another and that the concept of inter-actions among elementary particles as the basis of all energy was completely unknown.

Heat was considered a mysterious fluid; electricity was a mysterious fluid; friction, cohesion, and surface tension were mysterious forces that seemed to have no connection with kinetic or potential energy. Johann and Daniel Bernoulli of Switzerland were able to develop many uses for the idea of conservation of energy during the eighteenth century, especially in the study of fluid motion, but they were unable to make the jump required to apply this idea to other forms of energy.

As a result, the idea of conservation of energy was almost forgotten for many years, except that, as a practical matter, the French Academy formally resolved in 1775 to consider no more designs for perpetual-motion machines, because so much time had been wasted in discovering, over and over again, that these machines did not work.

At the end of the eighteenth century new ideas about energy began to emerge. Alessandro Volta's invention of the battery in 1800 showed that electric current could come from chemical reactions. The electric current could produce heat and light, and through magnetism it could produce motion. Motion, in turn, could produce electricity through friction. A closed cycle of energy conversion had thus been demonstrated: electricity, magnetism, mechanical motion, friction, electricity.

In 1822 the German physicist Thomas Johann Seebeck showed that heat applied to the junction between two different metals could produce an electric current directly. Twelve years later Jean Peltier of France reversed this, showing that a current applied to such a junction could produce heat or cold, depending on the direction of the current. For a time it was fashionable to write papers in the learned journals describing chains of transformations from one kind of energy to another, and people gradually began to speak of a single 'force' which could appear in electrical, thermal, dynamical, and other forms.

Earlier, in 1798, the American scientist Benjamin Thompson (who pursued a notorious career in Europe as Count Rumford) had shown that when cannons were bored out with a drill, the mechanical energy of the drill was converted into heat through friction. This discovery made little impact on the world of science until, in the years between 1840 and 1850, James Prescott Joule of England actually measured the amount of heat obtained from a given amount of mechanical energy. When it was demonstrated that mechanical work could be converted into heat without loss, the mental dam seemed to break, and all at once everybody was writing about conservation of energy. The brilliant French engineer Sadi Carnot had actually worked out the theory of the interchangeability of heat and mechanical work in the 1830s, but it took Joule's measurements of heat yields to attract wide attention and acceptance.

Between 1842 and 1847 the theory of conservation of energy was publicly announced by several scientists separately — by Joule, by Hermann von Helmholtz and Robert Mayer of Germany, and by others. It seemed that the idea was 'in the air' and just ripe for acceptance by the world of science. There has never been another such example of a theory appearing simultaneously in so many different minds.

There is good reason to believe this was no accident. In the first place, the beginning of the nineteenth century marked a great rise in engineering. People were becoming interested in water, wind, and steam power. The measurement of work and power became important. Use of the word 'work' became wide-spread in physics, and the equivalence of work (force times distance) and kinetic energy became well known. Between 1819 and 1839 several writers on engineering mechanics derived the expression $fd = \frac{1}{2}mv^2$.

Moreover, some of the German scientists – Helmholtz, Mayer, the chemist Justus von Liebig – had a concept of the indestructibility of energy even before they found evidence for it. When the evidence began to accumulate, they were prepared. They made the mental jump from the various discoveries to the general law of conservation of energy because they were looking for just such a principle.

The original influence behind all these simultaneous discoveries may have been a philosophical movement known in Germany as *Naturphilosophie*. Its adherents sought a single, unifying principle for all natural phenomena. The philosophy professor Friedrich Wilhelm Joseph von Schelling predicted in 1799 ' . . . that magnetic, electrical, chemical, and finally even organic phenomena would be interwoven into one great association . . . [which] extends over the whole of nature'. This is precisely the point of view that prevails today, many scientists believing that the universe will ultimately be explainable in terms of a few fields and elementary particles.

Schelling's followers dominated the teaching in German universities at the beginning of the nineteenth century, and the influence of this philosophy undoubtedly contributed to the recognition of conservation of energy when the facts were assembled.

*The Laws of Physics*, by M.A. Rothman

# 2 Heat

## Objectives

The intention of this chapter is to consider heat as a form of energy and the ways in which heat can be transferred from one object to another. Knowledge of the energy concept, the ideal gas laws (see *Book 2: Materials*) and electric circuits (see *Book 4: Basic Electricity and Magnetism*, Chapter 1) is needed. That part depending on the electricity could be delayed until basic electric circuits are considered.

The general objectives for this chapter are that after working through it you should be able to:

(a) Solve problems involving specific heat capacities and describe methods of measuring them;

(b) Calculate the temperature changes produced when electrical and mechanical energy are transformed into heat;

(c) Define molar heat capacity and state Dulong and Petit's law;

(d) Define the specific heat capacity at constant pressure and at constant volume and determine the difference between them for an ideal gas;

(e) Solve problems involving specific latent heats;

(f) Solve problems involving heat transfer by conduction through single- and multiple-layer structures;

(g) Distinguish between forced and natural convection;

(h) Solve problems involving the emission and absorption of radiation;

(i) State the zeroth law of thermodynamics;

(j) Explain how a temperature scale can be established.

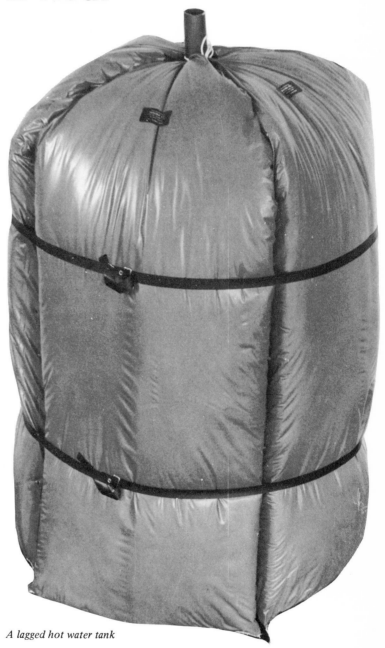

*A lagged hot water tank*

## Teaching note

Experiments appropriate to the first part of this chapter will be found in *Nuffield Physics: Text Year 4* and *Project Physics Course Handbook*. The second part of the chapter lends itself to experimental investigations, see *Physics Investigations* by W. Bolton

*Figure 2.1*

## Quantity of heat

If you put a pan of water over a Bunsen flame (*Figure 2.1*) then the temperature reached by the water depends on the length of time for which the water is heated. This suggests that the Bunsen flame is emitting a certain amount of energy per second and the temperature change depends on the total energy received, i.e., the quantity of heat. If you take readings with the heat emission of the Bunsen burner kept as near constant as possible, then the temperature rise will be found to be, within experimental error, proportional to the length of time. The temperature change seems to be proportional to the quantity of heat supplied.

Quantity of heat $\propto$ temperature change

A more accurate experiment using an electric immersion heater immersed in the water and with lagging to prevent heat losses to the surroundings confirms the proportionality.

If you repeat the above experiment with different amounts of water in the pan then the temperature rise per minute will be found to depend on the amount of water. The greater the amount of water present the smaller the temperature rise. A more careful experiment, with, say, the immersion heater and lagging around the vessel containing the water, will show that the temperature rise is inversely proportional to the mass of water present. Doubling the mass of water halves the temperature rise.

Temperature change $\propto 1/\text{mass of the water}$

for a constant quantity of heat supplied.

If the mass of water is doubled then to produce the same temperature change we need to double the quantity of heat supplied, perhaps by using the same level of the Bunsen flame but for twice the time. To produce double the temperature change with the same quantity of heat we can halve the mass of the water being heated.

Quantity of heat $\propto$ mass of water $\times$ temperature change

You should be able to see that this relationship fits the above facts by considering what happens when one of the quantities is constant and the other two change.

If another liquid, or a solid, is used in place of the water then the temperature change will be found to depend on the nature of the substance being heated.

Quantity of heat =
    specific heat capacity $\times$ mass $\times$ temperature change

The **specific heat capacity** depends on the material concerned. It can be defined as the quantity of heat needed to raise the temperature of 1 kg of the substance by 1 °C. The following are some typical values.

| Material | Specific heat capacity/J kg$^{-1}$ °C$^{-1}$ |
|---|---|
| Water, in liquid form | 4200 |
| Ice | 210 |
| Aluminium | 880 |
| Copper | 380 |
| Air, at atmospheric pressure | 1000 |

**Questions**  1  An electric immersion heater emitting a constant amount of heat per second is placed in 1 kg of water in an insulated container, and produces a temperature rise of 2 °C in 30 s.

(a) What temperature rise would the heater have produced in 30 s if placed in 2 kg of water?
(b) What temperature rise would the heater have produced in 2 minutes if placed in 1 kg of water?
(c) What is the quantity of heat being produced by the heater per second? This is the power of the heater.

2  What heat energy is needed to raise the temperature of 500 g of aluminium by 5 °C?

3  Is more, or less, heat energy needed to raise the temperature of a block of aluminium by 1 °C than to raise the temperature of a block of copper with the same mass by 1 °C?

4  Estimate the heat energy needed to raise the temperature of a room full of air by 5 °C.

In all the above calculations it has been assumed that all the heat energy supplied is used to increase the temperature of the object in question. This is rarely the case. Heat is generally lost to the surroundings or heat is received from the surroundings. Also, in the case of a liquid or a gas, heat has to be given to the container of the liquid or gas. The container has to be raised in temperature, the same as the liquid or gas.

**Questions**  5  An aluminium pan has a mass of 200 g and contains 1 kg of water.
What heat energy is needed to raise the temperature of the pan and the water by 30 °C?

6  If 30% of the heat energy supplied from the electric cooker to the pan and water in question 5 is used to raise the temperature of the pan and water and the remainder is lost to the surroundings, what heat energy must be supplied to raise the temperature of both the water and pan by 30 °C?

If 1 kg of water at a temperature of 50 °C is added to 1 kg of water at 30 °C what will be the resulting temperature? Each lot of water has had energy supplied to bring its temperature up to the value quoted and so the problem might be phrased as — what would be the result if all the energy supplied separately to the two lots of water had been supplied to just one lot with a mass of 2 kg, the combined mass? The result is 40 °C, you might check this experimentally.

Suppose we considered each lot of water to have been initially at 0 °C. Then the heat energy needed to raise the water to the quoted temperatures would have been

$$(4200 \times 1 \times 50) \text{ J} \quad \text{and} \quad (4200 \times 1 \times 30) \text{ J}$$

The total energy supplied would thus have been

$$(4200 \times 1 \times 50) + (4200 \times 1 \times 30) \text{ J}$$

If this energy had all been supplied to the combined mass of 2 kg, initially at 0 °C, then the temperature rise $\theta$ would have been given by

$$4200 \times 2 \times \theta = 4200 \times 1 \times 50 + 4200 \times 1 \times 30$$

Hence $\theta$ is 40 °C.

This analysis has assumed that heat is conserved. All we are doing is redistributing the heat. This is really just one form of the conservation of energy.

---

**Questions**   7   To 5 kg of water at 20 °C are added 2 kg of water at 60 °C. What will be the resulting temperature?

8   A block of copper having a mass of 100 g and at a temperature of 90 °C is dropped into 300 g of water at 20 °C. What will be the resulting temperature?

9   If the water in question **8** had been contained in a copper can having a mass of 200 g how would the resulting temperature have differed?

10   In an experiment to repeat question **9** it was found that the temperature resulting when the block of copper was dropped into the water in the copper can was lower than that obtained in the calculation. Why should this be the case?

---

## Heat from electrical energy

When a current passes through a wire, the wire becomes warm. The effect is evident in electric lamp bulbs where the wire becomes white hot, in the immersion heater used for hot water production in the home, in the electric fire used for heating the home. Electrical energy can be transformed into heat energy.

We can ascertain the factors that determine the heat energy produced by an electrical heater by doing an experiment with the heater immersed in a known mass of liquid, of known specific heat capacity, and measuring the temperature rise so produced (*Figure 2.2*). The heat energy produced is found to depend on the length of time for which the current is passed through the heater. The heat energy is directly proportional to the time. The heat energy is also found to depend on the current used and the voltage across the heating element; the energy is proportional to the product of the current and voltage

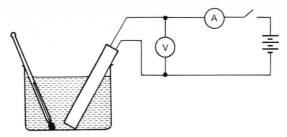

*Figure 2.2*

(see *Book 4: Basic Electricity and Magnetism* for a more detailed discussion). We can thus consider the heat energy to be related to *IVt*, where *I* is the current, *V* the voltage and *t* the time. By defining suitable units the relationship will be

Heat energy $= IVt$

This means that

Electrical energy $= IVt$

Units: energy − J, $I$ − amperes (A), $V$ − volts (V), $t$ − s.

The rate of transformation of energy is the power and so the electrical power must be

Electrical power $= IV$

Units: power − J/s or W, $I$ − A, $V$ − V.

**Questions**  **11**  How much heat energy will be produced per second by an electrical heater of power 50 W?

**12**  A current of 4 A passes through an electrical heater immersed in a liquid of specific heat capacity 1600 J kg$^{-1}$ °C$^{-1}$. Calculate the rise in temperature of the liquid that takes place in 2 minutes. Assume the calorimeter containing the liquid has a mass of 150 g and a specific heat capacity of 380 J kg$^{-1}$ °C$^{-1}$, there is 100 g of the liquid, the voltage across the heating coil is 1.5 V and no heat is gained or lost by the system.

## Heat from friction

Rubbing your hands together is one way of generating heat by friction. Joule's experiment described in the previous chapter concerned the heat produced by friction between a paddle wheel and water. If you have ever used a drill to make holes in a piece of metal or a saw to cut metal you will have found that the friction between the drill or saw and the metal produced heat.

All the above examples involve moving an object through a distance against a force, the frictional force. The product of this force and the distance gives the work, i.e., the energy transfer.

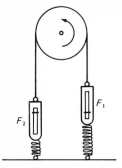

*Figure 2.3*

One method of measuring the power developed by a motor is to wrap a band brake round the motor shaft or flywheel, as in *Figure 2.3*. When the shaft rotates the brake rubs against it and the work done by the friction results in heat energy being produced. When the shaft is not rotating the spring balances on each end of the band indicate the same reading. When the shaft rotates the readings differ. This difference in reading is the frictional force as there is no net force causing any acceleration of the band.

$$\text{Frictional force} = F_1 - F_2$$

For one revolution of the shaft the distance moved against this frictional force will be equal to the circumference of the shaft, i.e., $2\pi R$, where $R$ is the shaft radius. If the shaft rotates at $f$ revolutions per second then the distance covered in 1 s will be $2\pi Rf$. Hence the work per second is given by

$$\text{Work per second} = (F_1 - F_2) \times 2\pi Rf$$

But the work per second is the power, hence

$$\text{Power} = 2\pi Rf(F_1 - F_2)$$

This power is the rate at which mechanical energy is transformed into heat energy.

**Question**   **13**   Calculate the power developed by a motor if a band brake wrapped around the motor flywheel shows a difference in tension between the two sides of the brake of 200 N when the flywheel rotates at 50 revolutions per second. The flywheel has a radius of 0.15 m.

### The measurement of specific heat capacity for solids and liquids

The measurement of specific heat capacity with any degree of accuracy is not easy. A simple method would be to take a block of the solid for which we need to know the specific heat capacity and to heat it up to some temperature which can be measured. You might, for instance, consider heating it in steam and assume that the block is at the temperature of the steam. The hot block would then be dropped into a lagged container holding, say, water. The temperature of the water would be measured before the hot block entered and then after the block had been 'mixed' with the water. We then might consider the situation to be:

Heat lost by the hot block in cooling to the 'mixture' temperature
= heat gained by the water and container in rising to that temperature.

If the mass and specific heat capacity of the water and the container are known the specific heat capacity of a block of known mass can be estimated. The problems with this method are determining the initial temperature of the hot block; knowing when the block, container and

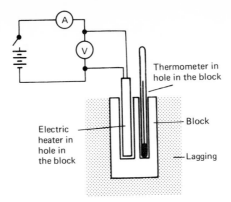

*Figure 2.4  A simple form of the electrical method for the measurement of specific heat capacity*

water have all reached the 'mixture' temperature; the measurement of this 'mixture' temperature and the problem that during the experiment there are likely to be heat exchanges between the block–water–container system and the surroundings. The above method is known as the **method of mixtures.**

Modern methods of measuring specific heat capacities tend to use an **electrical method**. An electrical heating element may be inserted into a block of the substance (*Figure 2.4*) for which the specific heat capacity is required, or wound round the block. The electrical power supplied to the heater can be determined from a measurement of the current and the voltage, power = current × voltage. The temperature of the block is measured both before and after the heater has been switched on. The time for which the heater is on is also measured. The energy used to heat the block is thus current × voltage × time. If there is no heat lost or gained from the surroundings then

Energy supplied by heater =
heat gained by the block plus any container

Liquids can have their specific heat capacities measured in a similar way.

Another method, known as a **continuous flow method**, was developed in 1899 by Callender and Barnes for the measurement of the specific heat capacity of water, and can be used for other liquids. *Figure 2.5* shows a simple form of this apparatus. The water is supplied from a

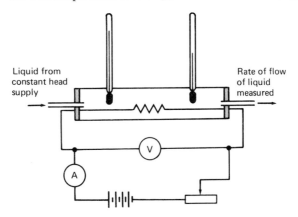

*Figure 2.5  A simple form of continuous flow calorimeter*

constant head supply so that it flows at a constant rate through the apparatus. The water flows through a tube which contains the electrical heating element along its axis. The temperature of the water entering and leaving this tube is measured, as also is the rate at which the water passes through the tube. The energy supplied to the heating element is determined from a measurement of the voltage and current. When the two thermometer readings become steady we can assume that all this supplied energy is going to heat the water and to some extent the surroundings. Under the steady conditions none of the energy is used to heat the apparatus since this is then at a constant temperature. Thus

Energy supplied by heater per second $=$
heat gained by water per second
$+$ any heat losses from the apparatus per second.

The above equation indicates one of the advantages of this method — we do not have to consider the heat given to the container. The heat gained by the water per second is equal to the mass of water flowing through the apparatus per second, $m$, multiplied by the specific heat capacity, $c$, and the difference in temperature indicated by the two thermometers, $\theta_2 - \theta_1$. Thus

$$IV = m \times c \times (\theta_2 - \theta_1) + h$$

where $h$ is the heat loss per second to the surroundings. While, by careful design of the apparatus, the heat loss to the surroundings is minimized there will inevitably be some loss. To eliminate this term the experiment is repeated with a different mass of water flowing through the apparatus per second but the electrical supply adjusted to give the same temperature difference. Then, because the apparatus is at the same temperature as in the first experiment, the heat losses can be expected to be the same and so

$$I'V' = m' \times c \times (\theta_2 - \theta_1) + h$$

Subtracting the two equations gives

$$IV - I'V' = (m - m') \times c \times (\theta_2 - \theta_1)$$

Hence the heat loss term is eliminated and the specific heat capacity can be determined.

---

**Question 14**   The following are the results obtained in a continuous flow calorimeter experiment.
What is the specific heat capacity of the liquid used?

|  | First experiment | Second experiment |
|---|---|---|
| Voltage across heating coil | 4.5 V | 6.0 V |
| Current in heating coil | 1.5 A | 2.0 A |
| Mass of water per second | 0.390 g | 0.750 g |
| Inlet temperature | 38.2 °C | 38.2 °C |
| Outlet temperature | 42.2 °C | 42.2 °C |

## Molar heat capacity

The specific heat capacity is the quantity of heat needed to raise the temperature of 1 kg of a substance by 1 °C. For aluminium this is 880 J kg$^{-1}$ °C$^{-1}$. One mole of aluminium has a mass of about 27 g. Thus the heat required to raise the temperature of one mole of aluminium by 1 °C is 27 × 10$^{-3}$ × 880 or about 24 J. This is referred to as the molar heat capacity of the aluminium. The **molar heat capacity** is the quantity of heat needed to raise the temperature of one mole of a substance by 1 °C.

For many solids the molar heat capacity is about 25 J mol$^{-1}$ °C$^{-1}$. This fact is called **Dulong and Petit's law.**

**Question 15** Copper has a specific heat capacity of 380 J kg$^{-1}$ °C$^{-1}$ and one mole has a mass of 63.5 g.
What is the molar heat capacity?

## Heat capacities of gases

The specific heat capacity of a solid or liquid is normally that measured under conditions of constant pressure, the constant pressure being that of the atmosphere. There is, however, another way in which we could have considered the specific heat and that is under conditions of constant volume. When heat is supplied to a substance and there is no change in volume then there is no work output and so all the supplied energy goes into raising the temperature of the substance. If, however, the heat is supplied when the pressure is kept constant then, because the volume of the substance increases, work is done and thus not all the supplied energy goes into raising the temperature.

**Question 16** Which would you expect to be larger for a substance – the specific heat at constant pressure or the specific heat at constant volume?

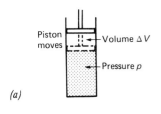

(a)

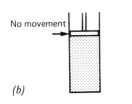

(b)

*Figure 2.6 (a) Constant pressure, (b) constant volume*

In the case of a gas the work done when a gas at pressure $p$ expands (*Figure 2.6*) by a small volume $\Delta V$ is $p\Delta V$ (*see* Chapter 1). Thus the change in internal energy $\Delta U$ that occurs when heat $Q$ is applied to a gas at constant pressure is

$$\Delta U = Q - p\Delta V$$

The **specific heat at constant pressure**, $c_p$, is the heat needed to raise the temperature of 1 kg of the substance at constant pressure by 1 °C. Thus

$$c_p = \frac{Q}{m \times \theta}$$

Hence

$$\Delta U = c_{\mathrm{p}} m \theta - p \Delta V$$

where $m$ is the mass of gas concerned and $\theta$ the change in temperature produced.

In the case of a gas at constant volume all the heat supplied goes into increasing the internal energy of the gas, no work being done. Thus

$$\Delta U = Q$$

The **specific heat at constant volume**, $c_{\mathrm{v}}$, is the heat needed to raise the temperature of 1 kg of the substance at constant volume by 1 °C. Thus

$$c_{\mathrm{v}} = \frac{Q}{m \times \theta}$$

Hence

$$\Delta U = c_{\mathrm{v}} m \theta$$

If we supply sufficient heat in both the constant pressure and constant volume situations to produce the same temperature rise, and hence the same internal energy change, then

$$\Delta U = c_{\mathrm{v}} m \theta = c_{\mathrm{p}} m \theta - p \Delta V$$

and so

$$c_{\mathrm{v}} = c_{\mathrm{p}} - \frac{p \Delta V}{m \theta}$$

For an ideal gas the temperature change $\theta$ produced by a volume change $\Delta V$ is given by

initially,

$$\frac{pV}{T} = mR \quad \text{or } pV = mRT$$

and after a temperature change $\theta$ and a volume change $\Delta V$,

$$\frac{p(V + \Delta V)}{(T + \theta)} = mR$$

or

$$pV + p\Delta V = mRT + mR\theta$$

and so

$$p\Delta V = mR\theta$$

This then gives for our specific heat equation

$$c_{\mathrm{v}} = c_{\mathrm{p}} - R$$

where $R$ is the characteristic gas constant. This is equal to the universal gas constant of 8.3143 J mol$^{-1}$ K$^{-1}$ divided by the molar mass (*see Book 2: Materials*).

If the molar heat capacities at constant volume and constant pressure, $C_v$ and $C_p$, are used then

$$C_v = C_p - R_0$$

where $R_0$ is the universal gas constant. The molar heat capacity at constant pressure of helium has been determined as 21.0 J mol$^{-1}$ K$^{-1}$ and that at constant volume as 12.6 J mol$^{-1}$ K$^{-1}$, a difference of 8.4 J mol$^{-1}$ K$^{-1}$ — in reasonable accord with the value of $R_0$.

**Questions**

**17**  The specific heat capacity of hydrogen ($H_2$) at constant pressure has been found to be $14.20 \times 10^3$ J kg$^{-1}$ K$^{-1}$.
What would you expect the specific heat capacity at constant volume to be?

**18**  The molar heat capacity of oxygen ($O_2$) at constant pressure is 29.1 J mol$^{-1}$ K$^{-1}$.
What would you expect the molar heat capacity at constant volume to be?

The ratio of the two specific heat capacities of a gas gives information about the atomicity of the gas.

$$\text{Ratio, } \gamma = \frac{\text{specific heat capacity at constant pressure}}{\text{specific heat capacity at constant volume}}$$

$$= \frac{\text{molar heat capacity at constant pressure}}{\text{molar heat capacity at constant volume}}$$

For a monatomic gas this ratio tends to have a value of about 1.67. For a diatomic gas it is about 1.40, for a polyatomic gas 1.30. You might like to check these values with the specific heat capacities given for hydrogen in question **17** or the molar heat capacities for oxygen in question **18**.

### Latent heat

*Figure 2.7* shows how the reading of a thermometer placed in ice changes as the ice is heated and melts. Though heat is supplied for the entire time the temperature does not continually increase. When the ice is below its melting point, 0 °C, the heat energy supplied causes the temperature of the ice to be increased. When, however, the ice is at its melting point the heat energy supplied does not result in any change of temperature. While the ice is melting the heat energy supplied does not give any change in temperature. When all the ice has melted the supply of heat energy once again causes an increase in temperature. The absence of a temperature change while the ice is melting indicates that heat energy is required to change the water from solid to liquid. The heat is known as the **latent heat of fusion**.

The amount of heat energy needed to change 1 kg of a solid at its melting point into liquid at the same temperature is called the **specific latent heat of fusion**. For water the specific latent heat of fusion is

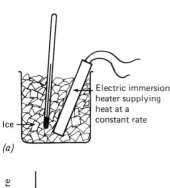

*(a)*

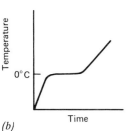

*(b)*

*Figure 2.7 (a) The experimental arrangement, (b) the experimental result*

$0.34 \times 10^6$ J/kg. Thus to melt a 2 kg block of ice, i.e., convert it from solid to liquid without any change of temperature, the heat energy required is $2 \times 0.34 \times 10^6$ J. The following are some typical specific latent heats of fusion.

| Material | Specific latent heat of fusion/MJ kg$^{-1}$ |
|---|---|
| Water | 0.34 |
| Aluminium | 0.39 |
| Lead | 0.025 |

Lead has a much smaller specific latent heat of fusion than aluminium. This means that for equal masses of the two substances much less heat energy is required to melt lead than aluminium.

**Question   19**   How much heat energy is required to melt a block of ice at 0 °C and which has a mass of 200 g?

When a liquid is cooled to such an extent that it freezes and becomes solid the specific latent heat of fusion is the amount of heat energy that has to be extracted from the liquid at its freezing point before it can change into the solid at the same temperature.

**Question   20**   How much energy has to be extracted from 20 kg of molten aluminium at its freezing point before it will solidify?

When water changes from liquid to vapour, i.e., steam, energy is required for the change. Thus if water is heated to 100 °C, then while it is changing from liquid to steam there is no change in temperature although heat energy is continually being supplied. The amount of heat energy needed to change 1 kg of liquid to vapour at the same temperature is called the **specific latent heat of vaporisation**. For water the specific latent heat of vaporisation is $2.27 \times 10^6$ J/kg. Thus to vaporise 2 kg of water, i.e., change it from liquid at 100 °C to steam at 100 °C, the heat energy required is $2 \times 2.27 \times 10^6$ J. The following are some typical specific latent heats of vaporisation.

| Material | Specific latent heat of vaporisation/MJ kg$^{-1}$ |
|---|---|
| Water | 2.27 |
| Alcohol | 0.86 |
| Mercury | 0.28 |

Alcohol has a smaller specific latent heat of vaporisation than water. This means that for equal masses of the two substances much less heat is required to vaporise the alcohol than the water; alcohol evaporates more readily than water.

**Question** 21 How much energy has to be extracted from 200 g of mercury at its liquefaction point before it will liquefy?

*Figure 2.8* gives graphs showing how the temperatures of water and alcohol change as the heat energy input into 1 kg of each substance is increased. In each case the heat energy inputs to or extractions from the substance are reckoned from an initial temperature of 0 °C. Thus in the case of water, as a liquid, at 0 °C an increase in the heat supplied to the water results in a rise in temperature until the water reaches 100 °C. The rate at which the temperature rises with heat energy input depends on the specific heat capacity. At 100 °C there has to be a large heat energy input to the water before the temperature starts to rise again. When this happens the water is a vapour, i.e., steam. If the liquid water is cooled below 0 °C there has to be a heat energy output to change the water from liquid to solid, i.e., ice, before the temperature starts to decrease. As will be clearly seen from the graph the specific latent heat of fusion is much less than that of vaporisation.

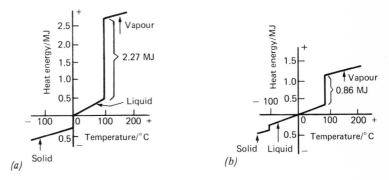

*Figure 2.8 Heat energy/temperature graphs for (a) water, (b) alcohol*

**Questions** 23 Is the specific latent heat of fusion of alcohol greater or less than that of water? Use the graphs in *Figure 2.8* to obtain your answer.

24 Is the normal boiling point of alcohol lower or higher than that of water? Use the graphs in *Figure 2.8* to obtain your answer.

25 Is the specific heat capacity of ice lower or higher than that of liquid water? Use the graph in *Figure 2.8(a)* to obtain your answer.

26 Which would be the better method of cooling a drink – adding an ice cube or adding an equal amount of liquid water at the same temperature as the ice? Explain your answer.

27    A cheap way of cooling water is to store it in a porous vessel. How does the evaporation of water through the pores of the vessel cool the water?

---

## Heat transfer

Conduction, convection and radiation are the three basic mechanisms by which heat is considered to be transferred between bodies. If you hold one end of a steel rod and the other end is in the fire it will not be very long before your end of the rod becomes hot. There has been a heat flow along the rod — from the high temperature end to the low temperature end. This form of heat transfer through a material with no apparent movement of the material is called **conduction**. If you stand a pan of water on the hot element of an electric cooker the water throughout the entire pan becomes warm although only the water at the bottom of the pan was heated. The water at the bottom of the pan is heated as a result of conduction through the metal at the base of the pan. This water then becomes warm and expands. The result of this expansion is that the density of the warmer water decreases and so the warmer water rises through the colder water. This movement of the water results in heat being transferred from the high temperature region to the lower temperature region by movement of the substance. This mode of heat transfer is called **convection**. The sun warms the earth, the heat energy being transferred through a vacuum. This mode of transfer without involving any intervening medium is called **radiation.**

## Thermal conductivity

The rate of transfer of heat energy by conduction through a block of material depends very much on the nature of the material concerned. Some materials are good conductors of heat, e.g., metals, while others are bad conductors, or insulators, e.g., the fibreglass blankets used for loft insulation or the blanket you put on your bed. Insulators are generally materials which contain pockets of trapped air. The air is the insulator and it is trapped so that it cannot give rise to convection currents and so cannot transfer the heat.

The rate at which heat energy is transferred through a material can be determined experimentally. *Figure 2.9* shows the basic form of an experiment. An electric heater supplies heat at a constant rate to one side of the block being investigated. At the other side of the block water is circulated past the end face. The rate at which heat is transmitted through the block, when the temperatures become steady, can be determined from a measurement of the increase in temperature of the water and its flow rate.

    Quantity of heat transmitted per second  =
                mass of water flowing per second
                × specific heat capacity of water
                × rise in temperature of the water due to the heat flow.

An alternative is to measure the electrical energy supplied to the electric heater per second. The entire arrangement is lagged, i.e., insulated, so

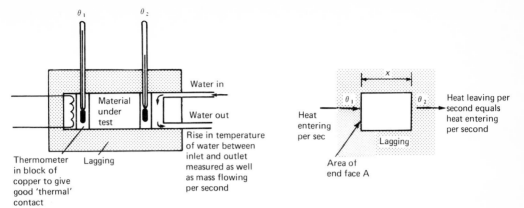

*Figure 2.9*

that all the heat energy from the heater passes through the block and out of the end face that is cooled by the water.

Experiment indicates that, when steady state conditions occur, there is a relationship between the rate of flow of heat through the block and the difference in temperature between the two end faces of the block — they are proportional when the block is well lagged.

Quantity of heat transmitted per second $\propto (\theta_2 - \theta_1)$

There is also a dependence on the size of the block, the quantity of heat being proportional to the cross-sectional area $A$ of the block and inversely proportional to the length $x$ of the block.

Quantity of heat transmitted per second $\propto \dfrac{A(\theta_2 - \theta_1)}{x}$

The other factor is the nature of the material concerned. All these facts can be represented in the equation

Quantity of heat transmitted per second $= kA \dfrac{(\theta_2 - \theta_1)}{x}$

where $k$ is a constant for a particular material and is known as the **thermal conductivity**. $(\theta_2 - \theta_1)/x$ is known as the temperature gradient.

The following are some typical values of thermal conductivities.

| *Material* | *Thermal conductivity*/J s$^{-1}$ m$^{-1}$ °C$^{-1}$ |
|---|---|
| Copper | 380 |
| Mild steel | 54 |
| Aluminium | 230 |
| Brick | 1 |
| Fibre-glass insulation | 0.04 |
| Water | 0.61 |
| Oil, as used in transformers | 0.13 |
| Air | 0.026 |

All the above are the values at room temperature, the values change quite significantly with temperature. In the case of the liquids and gases the values assume that there is no convection occurring. Thus the value for air is for 'trapped' air.

**Questions**   28   Which is a better conductor of heat, copper or aluminium?

29   The temperature outside a brick-constructed house is constant. How does the heat flow through the brick depend on the thickness of the brick if a constant inside temperature is maintained by a heater? Assume the walls of the house are constructed of just single layers of brick and also assume all the heat entering the internal face of the brick is conducted through to the external face.

30   Calculate the heat flow per second occurring through a well lagged bar of mild steel of length 400 mm and cross-sectional area 100 mm$^2$ if a temperature difference of 20 $^\circ$C exists between the ends of the bar.

In many situations the heat flow occurs through not just one material but through a number of layers of different materials. One example of this is the wall of the conventional house. Modern houses do not have an outer wall consisting of a single layer of brick but have what is called a cavity wall. This consists of two brick walls separated by a cavity. The heat from the house thus flows through a brick layer, an air layer and then another brick layer before it reaches the outside. There might even be another layer, the inside wall may be coated with a layer of plaster. *Figure 2.10(a)* shows the form of the double cavity wall. The reason for this construction is to reduce heat losses from the house.

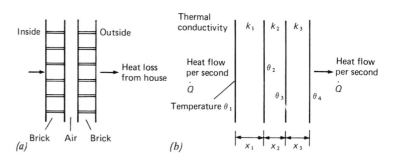

*Figure 2.10  The cavity wall*

If we consider the flow of heat through a multi-layer structure (*Figure 2.10(b)*) then at the steady-state condition:

*For the first layer*

$$\text{Heat flow per second } \dot{Q} = k_1 A \frac{(\theta_1 - \theta_2)}{x_1}$$

*For the second layer*

$$\text{Heat flow per second } \dot{Q} = k_2 A \frac{(\theta_2 - \theta_3)}{x_2}$$

*For the third layer*

Heat flow per second $\dot{Q} = k_3 A \dfrac{(\theta_3 - \theta_4)}{x_3}$

Rearranging these equations gives

$$\theta_1 - \theta_2 = \frac{\dot{Q}}{A} \frac{x_1}{k_1}$$

$$\theta_2 - \theta_3 = \frac{\dot{Q}}{A} \frac{x_2}{k_2}$$

$$\theta_3 - \theta_4 = \frac{\dot{Q}}{A} \frac{x_3}{k_3}$$

If these three equations are added together, then

$$\theta_1 - \theta_4 = \frac{\dot{Q}}{A} \left( \frac{x_1}{k_1} + \frac{x_2}{k_2} + \frac{x_3}{k_3} \right)$$

Rearranging this gives

$$\dot{Q} = A \frac{1}{\left( \dfrac{x_1}{k_1} + \dfrac{x_2}{k_2} + \dfrac{x_3}{k_3} \right)} (\theta_1 - \theta_4)$$

The rate of flow of heat $\dot{Q}$ thus depends on the temperature difference between the inside and the outside, the intermediate temperatures need not be known. This equation can be simplified to

$$\dot{Q} = UA\Delta\theta$$

where $\Delta\theta$ is the overall temperature difference and $U$ is a constant for that particular structure and is known as the **overall heat transfer coefficient** or sometimes just referred to as the *U* **value**.

$$U = \frac{1}{\left( \dfrac{x_1}{k_1} + \dfrac{x_2}{k_2} + \dfrac{x_3}{k_3} \right)}$$

In many instances the temperatures of the two surfaces of either a single wall or a composite wall are not known. In the case of a brick wall the temperatures generally known are those of the air in the room on one side and the surroundings on the other side. These are not the temperatures of the surfaces of the wall. Within the room there is probably a reasonably uniform temperature due to natural circulation of the air and convection. Close to the wall there is, however, an air layer that does not circulate. This layer is called the boundary layer and effectively 'sticks' to the wall owing to viscosity (*see* Chapter 4, *Book 2: Materials*). If you blow across any surface, e.g., a table top, you will find it impossible to blow away the very fine particles of dust on the surface. This is because the air layer adjacent to the surface has its velocity reduced by what we might call a frictional effect. The effect of this boundary layer on the transfer of heat through a wall is that we have to consider the boundary layer as another layer through which heat has to flow, mainly by conduction (*Figure 2.11*). The layers of air act as insulators on the two surfaces of the wall. Their effect can be taken into

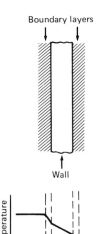

Boundary layers

Wall

Temperature

Distance

*Figure 2.11*

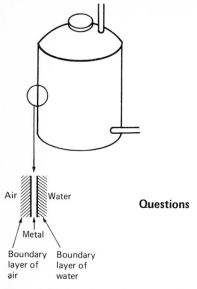

Air | Water

Metal

Boundary layer of air   Boundary layer of water

*Figure 2.12  An unlagged hot water tank*

account by including the effects of these extra layers in an overall heat transfer coefficient $U$.

$$\dot{Q} = UA\Delta\theta$$

$\Delta\theta$ is the temperature difference between the surroundings on either side of the wall.

Boundary layers occur wherever there is a fluid, i.e., a liquid or a gas, on one or both sides of a wall. In the case of the domestic hot water tank there is a boundary layer of hot water on the inside of the tank and a boundary layer of air on the outside (*Figure 2.12*).

**Questions**

31  The heat transfer coefficient for a cavity wall is 1.5 $\mathrm{J\,s^{-1}\,m^{-2}\,°C^{-}}$ What is the heat loss per second through a cavity wall of area 20 $\mathrm{m^2}$ when the temperature difference between the inside and outside is 14 °C?

32  A single glazed window has a heat transfer coefficient of 3.9 $\mathrm{J\,s^{-1}\,m^{-2}\,°C^{-1}}$. A double-glazed window has a heat transfer coefficient of 1.7 $\mathrm{J\,s^{-1}\,m^{-2}\,°C^{-1}}$.
For a window of area 2 $\mathrm{m^3}$ how does the use of double-glazing rather than single glazing reduce the heat loss when the difference in temperature between the inside and outside is (a) 5 °C and (b) 20 °C?

33  Plaster has a coefficient of thermal conductivity of 0.5 $\mathrm{J\,s^{-1}\,m^{-1}\,°C^{-1}}$ and brick 1.0 $\mathrm{J\,s^{-1}\,m^{-1}\,°C^{-1}}$. Calculate the heat loss per square metre through an internal wall in a house when the wall consists of a layer of plaster 15 mm thick on each side of brick of thickness 100 mm and the temperature difference between the two sides of the wall is 8 °C.
Why is this temperature difference not the same as that prevailing between the room temperatures on each side of the wall?

## Convection

Convection is an important method of transferring heat from one object to another. The conventional chemistry laboratory condenser (*Figure 2.13*) circulates water round a tube containing hot gases. The heat flows from the hot gases, through the wall of the condenser, to the water which moves and takes the heat away with it. This is a convection process. Convection is where the heat is conducted into a fluid which then moves away — whether the movement be due to the fluid rising naturally because its increase in temperature has meant a lower density or whether it be due to forced movement of the fluid by some fan or pump. These two forms of convection are known as **natural convection**

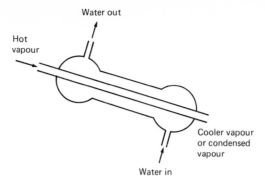

*Figure 2.13 Chemistry laboratory condenser, an example of forced convection*

and **forced convection** respectively. The laboratory condenser is an example of forced convection. The cup of coffee cooling on the table in a living room with no draughts is a reasonable example of cooling by natural convection.

For forced convection the rate of loss of heat $\dot{Q}$ per unit area of a surface is reasonably proportional to the temperature difference $\Delta\theta$ between the surface and its surroundings. This is known as **Newton's law of cooling.**

$$\dot{Q}/A \propto \Delta\theta$$

Thus, the greater the temperature difference between the surface and its surroundings the faster it loses heat. The equation can be written in the form

$$\dot{Q} = hA\Delta\theta$$

where $h$ is a heat transfer coefficient. The value of $h$ will depend on the properties of the fluid, i.e., density, thermal conductivity, specific heat capacity and viscosity, the fluid velocity and whether the fluid flow is streamline or turbulent.

Questions | 34 | How will the rate of loss of heat from a hot cup of coffee at 80 °C in a draught compare with that at 30 °C if the surroundings are at 20 °C?

35 | With the chemistry laboratory condenser shown in *Figure 2.14* more vapour is condensed when the water flow rate through the condenser is high than when it is low.
Explain this.

For natural convection the rate of loss of heat $\dot{Q}$ is found to be reasonably proportional to the temperature difference $\Delta\theta$ between the surface and its surroundings raised to the power 5/4.

$$\dot{Q} \propto \Delta\theta^{5/4}$$

**Questions**   **36**   How will the rate of loss of heat from a hot cup of coffee at 80 °C in still air compare with that at 30 °C if the surroundings are at 20 °C?

**37**   A large proportion of the heat lost from a house to the surroundings is by convection.
If the loss is predominantly by natural convection, i.e., no strong winds are blowing, how might you expect the heat losses from a house maintained at 24 °C to compare with the same house maintained at 18 °C when the external temperature is 4 °C?

## Radiation

On earth we receive heat energy from the sun. This heat energy passes through the vacuum that exists in the space between us and the sun. This mode of heat transfer is called radiation. When an eclipse of the sun occurs and the sun's light is obscured a drop in temperature on the earth also occurs at the same time. This suggests that light and the heat radiation may be propagated through space in much the same way.

Any hot object radiates heat energy — by hot we mean any object at a temperature above absolute zero. On the basis of experimental measurements made by John Tyndal the rate of emission of heat energy from an object at a temperature $T$ (on the Kelvin scale) was expressed by Josef Stefan in 1879 as

$$\dot{Q}/A = e\sigma T^4$$

where $A$ is the surface area of the object, $e$ a quantity called the **emissivity** of the surface and $\sigma$ a constant called **Stefan's constant** which has a value of $56.7 \times 10^{-9}$ J s$^{-1}$ m$^{-2}$ K$^{-4}$. The relationship was later derived theoretically by L. Boltzmann and is thus sometimes known as the **Stefan–Boltzmann equation**.

The surface of an object which is a perfect emitter emits the maximum amount of radiation at a particular temperature. For such an emitter the emissivity has a value of 1. The equation then becomes

$$\dot{Q}/A = \sigma T^4$$

The emissivity of a surface is expressed as the fraction of the radiation per second per unit area that would be emitted by a perfect emitter at the same temperature. For example, the emissivity of a sheet of clean, highly polished copper is about 0.03. This means that the sheet emits 3/100th of the radiation that would be emitted by the perfect emitter. Oxidised copper has an emissivity of 0.8 and thus emits 8/10th of the radiation that would be emitted by the perfect emitter. The emissivity depends on the nature of the surface concerned.

The perfect emitter is often called a **black body**. When radiation falls on a surface it may be reflected or absorbed. A black object is one that does not reflect light — that is why it appears as black. An object at some given temperature is both emitting radiation and absorbing radiation — the radiation emitted by its surroundings. When it is in equili-

brium with its surroundings then it emits radiation at the same rate
as it absorbs radiation. Hence a good emitter of radiation must be a good
absorber of radiation. A perfect emitter is thus a perfect absorber. A perfect absorber of light is a black surface and hence the term black body
for a perfect absorber or perfect emitter of heat radiation. A surface
painted with matt-black paint has an emissivity of 0.97 — a reasonably
practical approach to a black body. Polished mild steel has an emissivity
of about 0.07 — it is a far from good emitter of radiation, in fact it is a
good mirror. A good reflector is a poor emitter.

If a small body of emissivity $e$ is completely surrounded by walls at
a temperature $T_s$, then the rate of absorption of radiation $\dot{Q}_s$ by the
body is given by (*Figure 2.14*)

$$\dot{Q}_s/A = e\sigma T_s^4$$

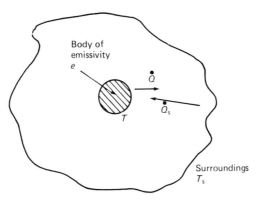

Figure 2.14 *Net rate of flow* $= \dot{Q} - \dot{Q}_s$

If the body itself is at a temperature $T$ then it is radiating at a rate $\dot{Q}$
given by

$$\dot{Q}/A = e\sigma T^4$$

The net rate of loss of heat $\dot{Q}_{net}$ by the body is thus

$$\dot{Q}_{net} = \dot{Q} - \dot{Q}_s$$

hence

$$\dot{Q}_{net}/A = e\sigma T^4 - e\sigma T_s^4$$
$$\dot{Q}_{net}/A = e\sigma(T^4 - T_s^4)$$

---

**Questions** 38  A domestic hot water tank is made of copper. In its normal state
this copper is oxidised and has an emissivity of about 0.8.
Which would reduce the heat loss from the tank — painting it
with gloss white paint or aluminium paint? Gloss white paint has
an emissivity of 0.9 and aluminium paint an emissivity of about 0.4.

39  Would polishing a surface be expected to lead to greater or less
heat loss from the surface?

**40**    A domestic hot water tank is a cylinder 1 m high and 0.4 m diameter. When full of hot water the surface temperature of the tank is 80 °C, the surroundings being at a temperature of 20 °C.
What is the rate of heat loss from the tank due to radiation if the tank surface has an emissivity of 0.8?

## The concept of temperature

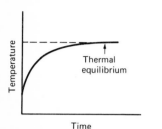

Figure 2.15 *The change with time of the reading given by a thermometer when placed in thermal contact with a hot object*

We can talk of measuring temperature with a mercury-in-glass thermometer. To do this we put the thermometer in close contact with the object for which the temperature is required and then wait for the reading given by the thermometer to become steady (*Figure 2.15*). When the steady state occurs we conclude that the thermometer and the object are at the same temperature.

The above argument has postulated the existence of a property called **temperature**. When two objects are put in thermal contact such that heat can flow between them then equilibrium will occur when the two objects reach the same temperature. The temperature of an object is that property which determines whether or not it will be in thermal equilibrium with another object or objects. When two or more systems are in thermal equilibrium they are said to be at the same temperature.

Two systems each of which is in thermal equilibrium with a third system are in thermal equilibrium with each other. This postulate is called the **zeroth law of thermodynamics**.

## A temperature scale

We use the simple mercury-in-glass thermometer as a temperature measurement system. This system depends on the variation in the length of a column of mercury with temperature. Temperature measurement systems depend on some material property varying with temperature. Thus the resistance thermometer depends on the variation of the electrical resistance of a coil of wire or semiconductor element with temperature. The gas thermometer gives a variation in pressure or volume with temperature.

When we use the mercury-in-glass thermometer it is put into thermal contact with the object for which the temperature is required and we then wait for thermal equilibrium to be reached. When this has occurred we read off the temperature from some scale which has values relating to the length of the mercury column. In order to give a temperature a value we thus need a temperature scale.

When a change of state occurs, e.g., solid to liquid, there is no change in temperature during the transition. It is possible, therefore, to take two change of state transitions and define their temperatures in terms of some arbitrary numbers. Between the year 1714 and 1717, D.G. Fahrenheit proposed a scale of temperature, known as the **Fahrenheit scale**, in which he called the freezing point of water 32 ° and the boiling point of water 212 °. During the period 1710 to 1743 another scale evolved, the **Celsius scale**. With this scale the freezing point of water was called 0 ° and the boiling point of water 100 °. Between the fixed points we might state that the change in temperature is proportional to the change in the thermometric property, e.g., the length of the mercury in the mercury-in-glass thermometer. Thus, if we immerse such a thermometer in melting ice we can put a mark on the thermometer for the mercury level that corresponds to the freezing point. Similarly, we can immerse the thermometer in the steam arising from boiling water (at normal atmospheric pressure as the boiling point depends on pressure) and mark the mercury level for the boiling point temperature. If we call these temperatures 0 ° and 100 ° then we can define 50 ° as being half way between the 0 ° and 100 ° marks, 25 ° a quarter of the way up the scale, etc. We have thus defined a temperature scale. *Figure 2.16* shows this graphically.

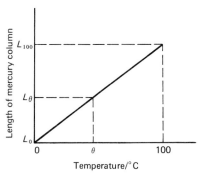

Slope of graph $= \dfrac{L_{100} - L_0}{100}$

hence length $L_\theta$ at temperature $\theta$ is given by

$$\frac{L_\theta - L_0}{\theta} = \frac{L_{100} - L_0}{100}$$

or

$$\theta = \frac{(L_\theta - L_0)}{(L_{100} - L_0)} \times 100\,^\circ C$$

*Figure 2.16*

The problem associated with defining a scale in the above way is that it depends on a particular property of a particular material and thus temperature scales derived by using different materials or different properties may not be the same. We can specify that the expansion of mercury in glass varies uniformly with temperature but then some other property might not vary uniformly with temperature according to this temperature scale.

The **International Practical Scale** of temperature is an internationally agreed scale which specifies a number of fixed points and the type of thermometers to be used to establish the temperatures between the points. Thus, for example, the temperature of equilibrium between ice, liquid water and its vapour (known as the **triple point** of water) is defined as being +0.01 °C and the temperature of equilibrium between liquid water and its vapour (the boiling point of water) is defined as being 100 °C under certain specified pressure conditions. The temperatures between these two fixed points are specified by means of a platinum resistance thermometer according to the formula $R_\theta = R_0(1 + A\theta + B\theta^2)$, where $R_\theta$ is the resistance of the platinum wire at temperature $\theta$ and $R_0$ the resistance at 0 °C. A and B are constants. The purity of the platinum used is closely specified. This equation means that the temperatures between the fixed points are not defined by means of a linear relationship, the resistance of the wire at 50 °C is not halfway between the resistances at 0 °C and 100 °C.

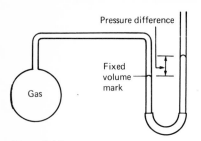

Pressure difference

Fixed
volume
mark

Gas

*Figure 2.17*

**Questions**

41   What are the advantages of specifying a temperature fixed point
by a change of state for a specific substance rather than in terms
of some property of a substance that varies with temperature,
e.g., the expansion of a piece of metal?

42   *Figure 2.17* shows a simple form of gas thermometer. The level
of the mercury in the manometer is adjusted by raising or lower-
ing the open limb of the U so that the volume of the gas in the
bulb remains constant whatever its temperature. The difference
in mercury levels gives the pressure difference between that of
the atmosphere and the gas in the bulb. How could you establish
a temperature scale for such a thermometer?

The fundamental scale that is used for temperature is the **thermo-
dynamic Kelvin scale**, temperatures on this scale being designated as K.
The size of the degrees Kelvin have been defined by the decision to
specify the triple point of water as being exactly 273.16 K. A degree
Kelvin is essentially the same as a degree Celsius and thus temperatures
on the Celsius scale can be converted to Kelvin scale temperatures by
adding 273.15 degrees to the Celsius value. Hence the boiling point of
water is 373.15 K. The origin of the Kelvin scale is indicated in
Chapter 3.

**Question**

43   What are the following temperatures on the Kelvin scale?
(a) 30 °C, (b) 200 °C, (c) −50 °C.

**Suggestions for answers**

1   (a) 1 °C
(b) 8 °C
(c) $4200 \times 1 \times 2/30 = 280$ J

2   $880 \times 0.500 \times 5 = 2200$ J

3   Aluminium needs more energy

4   Suppose the room to be 4 m × 4 m × 3 m, i.e. a volume of 48 m³
The density of air is about 1.2 kg/m³, hence the mass of air in the
room is about 57.6 kg. Hence the heat required is
$1000 \times 57.6 \times 5$ or 288 000 J.

5   $880 \times 0.2 \times 30 + 4200 \times 1 \times 30 = 131\ 280$ J

6   438 000 J

7   31.4 °C

8   $0.100 \times 380 \times 90 + 0.300 \times 4200 \times 20 = 0.100 \times 380 \times \theta$
$+ 0.300 \times 4200 \times \theta$. Hence $\theta$ is about 22.4 °C.

**9** $0.100 \times 380 \times 90 + 0.300 \times 4200 \times 20 + 0.200 \times 380 \times 20 = 0.100 \times 380 \times \theta + 0.300 \times 4200 \times \theta + 0.200 \times 380 \times \theta$. Hence $\theta$ is about 22.2 °C.

**10** Some heat is lost to the surroundings.

**11** 50 J

**12** $4 \times 1.5 \times 120 = 0.150 \times 380 \times \theta + 0.100 \times 1600 \times \theta$, hence $\theta$ is about 3.3 °C.

**13** $2\pi \times 0.15 \times 50 \times 200$ or about 9.4 kW

**14** $6.0 \times 2.0 - 4.5 \times 1.5 = (0.750 - 0.390) \times 10^{-3} \times c \times (42.2 - 38.2)$ This gives $c$ as about 3860 J kg$^{-1}$ °C$^{-1}$.

**15** $63.5 \times 10^{-3} \times 380$ or about 24 J mol$^{-1}$ °C$^{-1}$

**16** The specific heat capacity at constant pressure is greater as more heat has to be supplied to produce a particular temperature rise.

**17** $14\,200 - 8.3143/(2 \times 10^{-3})$ or about 10 040 J kg$^{-1}$ K$^{-1}$

**18** $29.1 - 8.4 = 20.7$ J mol$^{-1}$ K$^{-1}$

**19** $0.200 \times 0.34 \times 10^6 = 68$ kJ

**20** $20 \times 0.39 \times 10^6 = 7.8$ MJ

**21** $0.100 \times 2.27 \times 10^6 = 0.227$ MJ

**22** $0.200 \times 0.28 \times 10^6 = 56$ kJ

**23** Less. The step in the graph where the solid changes to liquid without any change in temperature is smaller for alcohol than water.

**24** Lower

**25** Lower, the slope of the graph is less, i.e., the heat energy per degree for a kilogram is less.

**26** Adding an ice cube. When the ice melts it extracts the latent heat from the drink. It is then at the same temperature as the added water so the latent heat extracted from the drink is a bonus.

**27** Evaporation is a change from liquid to vapour and involves heat being extracted from the water in the store – hence the water cools.

**28** Copper

**29** The greater the thickness the less the heat flow. $Q \propto 1/x$.

**30**   $54 \times 100 \times 10^{-6} \times 20/0.400 = 0.27$ J

**31**   $1.5 \times 20 \times 14 = 420$ J/s

**32**   Single glazing: $\dot{Q} =$ (a) 39 J/s, (b) 156 J/s
Double glazing: $\dot{Q} =$ (a) 17 J/s, (b) 68 J/s
Hence reduction for 5 °C is 22 J/s, for 20 °C it is 88 J/s

**33**   $U = \dfrac{1}{\left(\dfrac{0.015}{0.5} + \dfrac{0.100}{1.0} + \dfrac{0.015}{0.5}\right)} = \dfrac{1}{0.16}$ J s$^{-1}$ m$^{-2}$ °C$^{-1}$

$\dot{Q} = \dfrac{1}{0.16} \times 1 \times 8 = 50$ J/s

The $U$ value neglects the effects of the air layers clinging to each side of the wall and hence the calculation will not relate to the temperature difference between the surroundings on each side of the wall.

**34**   $\dfrac{80 - 20}{30 - 20} = 6$ times more

**35**   There is a greater temperature difference between the vapour and the water in the condenser because the faster flow keeps the water in the condenser cooler.

**36**   $\dfrac{(80 - 20)^{5/4}}{(30 - 20)^{5/4}} = 6^{5/4}$ or about 9.4

**37**   $\dfrac{20^{5/4}}{14^{5/4}}$ or about 1.57

The moral of this is that reducing the temperature at which you maintain your home can quite significantly reduce the heating bill.

**38**   Aluminium paint would reduce the heat loss by radiation.

**39**   Less heat being lost by radiation

**40**   $A = 1 \times \pi \times 0.4 + 2 \times \pi \times 0.4^2 /4$ or about 1.51 m$^2$
$\dot{Q} = 1.51 \times 0.8 \times 56.7 \times 10^{-9} \times (353^4 - 293^4)$ or about 555 J/s

**41**   No change in temperature occurs during the transition and so it is a clearly defined temperature. It can be easily and reproducibly realized at many parts of the world from a specification of the substance concerned. If the expansion was specified then not only would the material need to be specified but the unit of length would need specification.

**42**   Immerse the bulb in melting ice for the 0 °C point and then in steam for the 100 °C point.

**43**   (a) 303.15 K, (b) 573.15 K, (c) 223.15 K.

**Further problems**   No suggestions for answers are given for these problems.

**44**   What mass of water would be heated from 20 °C to 60 °C in 12 minutes by an electric immersion heater dissipating energy at 200 J/s?

**45**   An aluminium pan of mass 250 g contains 800 g of water. What heat energy is needed to raise the temperature of the pan and its contents from 20 °C to 100 °C?

**46**   What is the resulting temperature when 2 kg of water at 80 °C are mixed with 1 kg of water at 15 °C?

**47**   A cup of coffee has been made with hot water. Cold milk is then added to the black coffee in the cup.
Explain how you could calculate the temperature of the resulting white coffee if it is assumed that there are no heat losses from the cup or the coffee.

**48**   A metal teapot has a mass of 400 g and is at a temperature of 15 °C when 500 g of boiling water are poured into it.
What will be the resulting temperature of the water in the teapot if no heat losses to the surroundings occur? The metal has a specific heat capacity of 880 J kg$^{-1}$ °C$^{-1}$.

**49**   Describe a method by which you could measure the specific heat capacity of copper.

**50**   Explain the term latent heat.

**51**   Calculate the energy needed when 2 kg of ice at 0 °C changes to water at the same temperature.

**52**   Calculate the energy given out when 400 g of water at 0 °C changes to ice at the same temperature.

**53**   A cardboard cup, of negligible mass, contains 400 g of water at a temperature of 50 °C. Pieces of ice amounting to a mass of 20 g are added to the water.
What will be the resulting temperature if the ice was initially at 0 °C? Assume no heat is lost or gained by the system.

**54**   An ice cube of mass 40 g is taken from a deep freeze cabinet where its temperature was −20 °C and is dropped into water at 0 °C.
How much water will freeze onto the cube if no heat is gained or lost by the system? The specific heat capacity of the ice is 2000 J kg$^{-1}$ °C$^{-1}$.

**55**   How much heat energy is needed to take 300 g of water at 20 °C and convert it into steam at 100 °C?

**56**   An alcohol has a boiling point under normal conditions of 78 °C, a specific heat capacity of 2500 J kg$^{-1}$ °C$^{-1}$ and a specific latent heat of vaporization of 0.86 × 10$^6$ J kg$^{-1}$. Water under normal conditions has a boiling point of 100 °C, a specific heat capacity of 4200 J kg$^{-1}$ °C$^{-1}$ and a specific latent heat of vaporisation of 2.27 × 10$^6$ J kg$^{-1}$

(a)  How much energy is needed to take 50 g of alcohol at 20 °C and convert it to vapour at its boiling point?
(b)  How much energy is needed to take 40 g of water at 20 °C and convert it to vapour at its boiling point?
(c)  An electric immersion heater dissipating 50 J/s is placed in a mixture of 50 g of alcohol and 40 g of water.
How would you expect the temperature to vary with time?

**57**   An open pan contains water which is boiling.
If 8 g of the water boil away every second what is the temperature of the lower face of the base of the pan? The base of the pan has an area of 300 cm$^2$, a thickness of 3 mm, and is made of aluminium.

**58**   Will water heat up more quickly in a copper kettle or an iron kettle if the kettles are the same size and the material is of the same thickness?

**59**   How much heat is conducted per second through a lagged copper bar of length 200 mm and cross-sectional area 400 mm$^2$ if the temperature difference between the ends is 60 °C?

**60**   Electrical transformers become hot when in use.
What would you expect to be the effect on the temperature of the transformer of immersing it in oil rather than leaving it in air?

**61**   A refrigerator consists of an outer thin metal sheet, mild steel, and an inner thin metal sheet of the same metal with a layer of fibre-glass between. The total surface area of a refrigerator is 3 m$^3$. How much heat will leak through the walls of the refrigerator per second if the outside temperature is 20 °C and the inside 2 °C? You can neglect the thin metal sheets and consider the walls to be essentially just the insulation which has a thickness of 30 mm.

**62**   A cavity wall consists of brick 100 mm deep, a 50 mm cavity and then another brick 100 mm deep. The inner brick is coated with plaster, 15 mm thick. The brick has a thermal conductivity of 0.90 J s$^{-1}$ m$^{-1}$ °C$^{-1}$, the air in the gap 0.03 J s$^{-1}$ m$^{-1}$ °C$^{-1}$ and the plaster 0.16 J s$^{-1}$ m$^{-1}$ °C$^{-1}$.
What is the overall heat transfer coefficient for the wall?

**63**   A partition wall in a house consists of plasterboard on timber framing. If this has a heat transfer coefficient of 1.7 J s$^{-1}$ m$^{-1}$ °C$^{-1}$ what will be the heat flow per second through a square metre of such a wall when the temperature difference between the rooms on either side of the partition is 10 °C?

64 Insulating the loft of a house with an 80 mm thick layer of fibre-glass changes the heat transfer coefficient of the roof from 1.4 to 0.4.
How will this affect the heat loss per second through each square metre of roof?

65 A wool blanket has a thermal conductivity of 0.04 $J\,s^{-1}\,m^{-1}\,{}^{\circ}C^{-1}$. What is the rate of loss of heat per square metre through a wool blanket 10 mm thick when the temperature under the blanket is 23 $^{\circ}$C and above the blanket 12 $^{\circ}$C.

66 Explain how you would devise an experimental arrangement to measure the thermal conductivity of a wool blanket.

67 Explain how you would devise an experimental arrangement to measure the overall heat transfer coefficient of a double-glazed window.

68 Heat is produced in the cylinders of a car engine due to the combustion of the petrol. The engine is kept cool by water from the radiator being pumped through the engine. The radiator is kept cool by air passing through the grill in front of the radiator as a result of the motion of both the car and the fan mounted behind the radiator.
Explain the different heat transfer processes involved in this arrangement.

69 The heat transfer coefficient for forced convection over a smooth plate of length about 0.5 m where the fluid velocity is less than 5 m/s is given by $h = 0.005 + 0.004v$ where $v$ is the velocity in m/s and $h$ is in units of $kJ\,s^{-1}\,m^{-2}\,{}^{\circ}C^{-1}$.
What is the heat loss by forced convection from a smooth plate of length 0.5 m and width 0.4 m when air blows over the plate at 3 m/s and the plate has a temperature of 80 $^{\circ}$C above its surroundings?

70 Estimate the heat loss by radiation from a tank 2 m $\times$ 2 m $\times$ 1 m high when it is at a temperature of 50 $^{\circ}$C and the surroundings are at 15 $^{\circ}$C. The emissivity of the tank surface is 0.8.

71 Why do electric fire elements generally have a shiny reflector behind them? Would you expect the reflector to become hot?

72 A filament of a valve needs to dissipate 10 J per second by radiation.
What must the surface area of the filament be if the temperature of the filament is not to exceed 1500 K, the surroundings being considered to be at 300 K?

73 The earth receives energy by radiation from the sun at a rate of 1400 J per second per square metre of the earth. (This is known as the solar constant.) The distance of the sun from the earth is $1.5 \times 10^{11}$ m.

(a) What is the total radiation per second emitted by the sun?
(b) The sun has a diameter of $1.4 \times 10^9$ m. If the sun is assumed to be a black body what is its surface temperature?

74 A peephole in the wall of a furnace emits radiation at the rate of 100 J per second.
If the peephole has an area of 200 mm$^2$, what is the temperature of the furnace if it is assumed to be a black body?

75 Give the experimental details of how you would establish a scale on an ungraduated thermometer.

76 A temperature scale can be specified in terms of the behaviour of a 'permanent' gas when subject to temperature changes. Why should such a scale have advantages over a scale specified in terms of the expansion of liquids?

---

## Appendix 2.1 Energy saving in the home

According to figures issued by the Department of Energy in Britain for a typical semi-detached house which has not been insulated or draught-proofed, 35% of the heat loss from the house occurs through the walls, 25% through the roof, 15% into the ground, 15% in draughts and 10% through the windows. A 75 mm thick layer of glass or mineral fibre roll laid between the ceiling joists in the loft of the house can reduce the heat loss through the roof by up to 80%. Urea formaldehyde foam injected into the cavity of the brick outer wall can reduce the heat loss through the walls by about 65%. By careful insulation and draught-proofing, up to 64% of the heat loss from the house can be saved. Saving heat means not only lower heating bills for the house tenant but an energy saving for the country. This can mean that the world's energy resources might last a bit longer (see Chapter 4).

# 3 Engines

## Objectives

The intention of this chapter is to consider the concepts of energy and power in a technological situation — the development of engines. Chapter 1 of this book and Newton's laws (*Book 1: Motion and Force*) are assumed to have been covered. This chapter lends itself to a consideration in conjunction with a historical or economic study of the British industrial revolution.

The general objectives for this chapter are that after working through it you should be able to:

(a)  Explain the principles behind the Newcomen engine and the Watt engine and explain how the developments led to greater efficiency and power;

(b)  Explain the principles behind the reaction and impulse steam turbines;

(c)  Explain the principles behind the internal combustion engine;

(d)  Define thermal efficiency and recognise the relationship between the efficiency and the temperatures of the hot reservoir acting as the source of the heat and the cold reservoir into which the waste heat is exhausted.

*The new 'O' series engine in Princess 2*

## Teaching note

There are many sources of background reading, e.g., *A simple history of the steam engine* by J.D. Storey, *Heat engines* by J.F. Sandfort and *Project Physics Course, Text*. Though not concerned with steam but water, *Project Technology Handbook No. 11: Industrial archaeology of watermills and waterpower* is worth reading, particularly for its suggestions for investigations.

## Heat engines

The first practical steam engine was produced by Thomas Savery as long ago as 1698. Prior to this most work was done by man or animals, with some additional power coming from windmills and watermills. Windmills could not always be relied on and both windmills and watermills needed to be built in special places — on tops of hills or alongside streams or rivers. About the time at which the steam engine was produced there was a need for pumps to pump water out of coal mines so that coal other than that located near the surface could be extracted. It was this need which led directly to the invention of the steam engine — initially it was an engine to pump water out of mines and only later a general industrial source of power.

Savery's steam engine became commercially available in about 1702. By 1712, however, a new engine had been developed which was to replace Savery's engine and begin a line of development which leads almost up to today. The engine was developed by Thomas Newcomen. *Figure 3.1* shows a simplified diagram of the engine; it was a huge contraption, about ten metres high. Above the engine was a huge wooden beam pivoted at its centre and rocking up and down as a result of a piston moving in a cylinder. The purpose of the beam was to transfer power from the piston moving in the cylinder to the load or pump.

The load end of the rocking beam was heavier than the piston end. Thus, the normal position of the beam resulted in the piston being at the top of its cylinder. With the piston in this position valve A was opened to allow steam into the cylinder. When the cylinder was full of

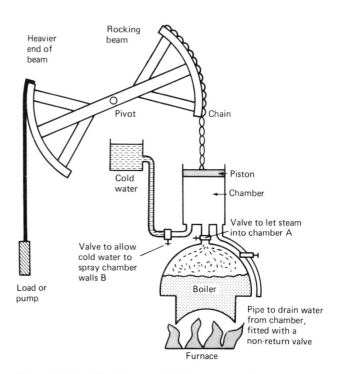

*Figure 3.1  Simplified diagram of a Newcomen engine*

steam the valve was closed. Valve B was then opened and allowed a spray of cold water to enter the chamber and condense the steam. The condensation of the steam led to a partial vacuum developing in the chamber below the piston. This produced a pressure difference between the atmospheric pressure on one side of the piston and the partial vacuum on the other side. The piston was thus pushed downwards. This movement caused the load end of the rocking beam to be raised. When the piston neared the bottom of the cylinder valve A was opened again and the heavier end of the beam then caused the piston to be pulled back up the cylinder.

**Questions**  1   Why does the production of a partial vacuum under the piston result in the piston moving down the cylinder? Your answer must include the term pressure.

2   The atmospheric pressure is about $10^5$ N/m$^2$.
If a complete vacuum had been produced in the cylinder then the pressure difference between the two sides of the piston would have been about $10^5$ N/m$^2$. The Newcomen engine of 1712 had a piston 0.54 m in diameter and a cylinder 2.4 m long.
If the piston moved through this entire cylinder volume, about 0.55 m$^3$, and the pressure difference remained constant during the entire movement, what would be the work done? *Figure 3.2* shows a very simplified pressure/volume graph for the piston movement.

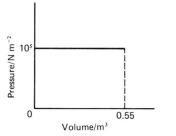

*Figure 3.2 A very simplified pressure/ volume graph for the Newcomen engine*

3   The Newcomen engine of 1712 was able to lift 46 kg of water through 46 m on each stroke of the piston.
What is the potential energy given to the water? How does this answer compare with that of the previous question? Would you expect any comparison?

4   The engine made one stroke, one cycle, every 5 s. Use your answer to question 3 to calculate the power of the engine.

The Newcomen engine was a heavy consumer of coal because every time the cylinder was filled with steam the cylinder itself had to be heated up to steam temperature. The overall efficiency of the engine was only about 2% as only about 12% of the steam was available as steam in the cylinder, the rest condensed on the walls of the cylinder.

To obtain greater power Newcomen made the cylinders bigger. Thus the volume swept through by the piston in a stroke was increased and hence the work done per stroke was increased. Making the cylinder bigger did, however, mean the cylinder had to have larger walls and because they were larger they also had to be thicker. This meant more metal had to be heated up by the steam condensing on it and so a lower efficiency for the engine.

**Question**  5  Newcomen built an engine in 1739 that had a piston of diameter 0.76 m and a stroke length, i.e., piston movement, of 2.7 m. The engine made 15 strokes every minute. What was the power?

It was James Watt who realized that the Newcomen engine was wasting steam in heating up the walls of the cylinder on every stroke. He arranged for the steam to be condensed in a condenser separate from the cylinder. He also used the steam to push the piston down rather than using atmospheric pressure for the down stroke. *Figure 3.3* shows the basic arrangement. The steam enters the top of the chamber and pushes the piston down. When the pressure of the steam above the piston has risen to a value sufficiently greater than the pressure below the piston the valve in the bypass arm opens and the steam can then pass into the space below the piston. The pressures on each side of the piston then tend to equalize. As this happens the piston is pulled back up the chamber by counterweights on the beam. When the piston has risen up the chamber a pump comes into operation to extract the steam from below the piston and condense it in a separate condenser. The entire cycle then repeats itself. During the entire operation the chamber is kept at steam temperature.

A later design by Watt involved applying the steam first to one side and then to the other side of the piston, so producing what is known

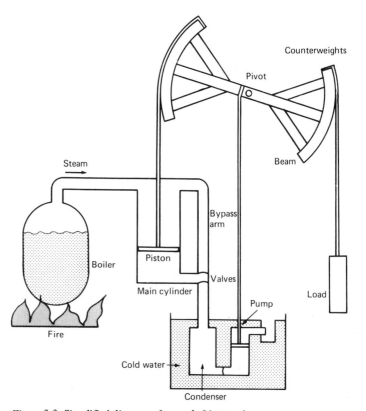

*Figure 3.3 Simplified diagram of an early Watt engine*

as a **double-acting engine**, i.e., using the steam to both push and pull instead of just pushing.

Newcomen's engines had overall efficiencies of about 2%, Watt's early engines gave about 3% and his later engines 4%. This was double the efficiency of Newcomen's engine but still very inefficient. This was some fifty or more years after Newcomen first produced his engine. Watt's early engines had powers of about 10 kW and his later engines powers of about 30 kW.

One innovation that did not start with Watt was the use of high pressure steam. This came from the work of O. Evans in 1802 and R. Trevithick, about 1800, who produced **high pressure engines**. At normal atmospheric pressure water boils at 100 °C, the steam produced thus being at atmospheric pressure, i.e., about $10^5$ N/m². To produce steam at a pressure of $2 \times 10^5$ N/m² the water has to boil at 121 °C. At a pressure of $6 \times 10^5$ N/m² the boiling point is 159 °C. The higher pressures thus meant higher boiler temperatures. Higher pressures also meant a greater chance of explosion − many of the early high pressure engines did explode. The higher pressures did mean that the pistons received higher thrusts from the steam and so higher stroke rates were possible. This meant that for the same volume chamber it was possible to achieve higher power.

Higher pressure engines were thus able to be smaller and still develop high power. The engines were lighter because of their smaller size. It was this that led to the development of the steam locomotive. By 1804 Trevithick had a small locomotive pulling coal wagons along rails. The person who is credited with being the founder of the railways is, however, G. Stephenson. Probably his most famous steam locomotive was the Rocket (*Figure 3.4*) which in 1829 won the prize for the best locomotive, covering almost 20 km in 54 minutes. The modern locomotive, no longer steam but electric or diesel−electric, achieves speeds of about 150 km/h.

By about 1830 engine efficiencies had risen to about 15% and powers of the order of 550 kW were being realised.

*Figure 3.4 Stephenson's locomotive Rocket*

## The steam turbine

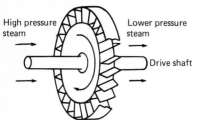

High pressure steam

Lower pressure steam

Drive shaft

*Figure 3.5 A reaction turbine*

Although steam engines are no longer widely used steam still is. The present-day electric power stations are, in the main, steam driven. The steam is used in a steam turbine to drive the electric generators. The first commercial steam turbine was invented by Charles Parsons in 1884.

In the **Parsons' turbine** the high pressure steam was allowed to 'escape' to a low pressure region through the blades of the turbine with the result that the turbine rotated. *Figure 3.5* shows the main features of one stage of the turbine. After passing through one stage the steam has lost only part of its pressure and it continues through a number of stages losing pressure at each stage. All the stages are connected to the same drive shaft.

**Questions**

6    In a hosepipe with a nozzle we have a situation where the water in the pipe escapes from its high pressure into the lower pressure outside in the air. A force is experienced by anybody holding the hosepipe — the force being in the opposite direction to that followed by the water when it leaves the nozzle.
Explain how this reaction force occurs.

7    When steam passes through the blades of the turbine there is a change in steam velocity — the velocity increasing as a result of the drop in pressure. There is thus a change in momentum.
In what direction must the change in momentum act for the turbine to rotate? What design factor of the turbine affects the direction of the change in momentum?

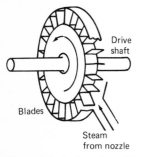

Drive shaft

Blades

Steam from nozzle

*Figure 3.6 An impulse turbine*

Parsons' first turbine produced a power of about 7.5 kW with the turbine rotating at 300 revolutions per second. The steam was at a pressure of more than five times normal atmospheric pressure. The efficiency was almost 20%.

In 1889 de Laval introduced another form of steam turbine, known as the **impulse turbine** (*Figure 3.6*). In this turbine the steam issues from a jet and directly strikes the turbine blades — rather like water hitting the blades on a water wheel. Both the de Laval and the Parsons forms of turbine are still in current use.

**Question**

8    When a high velocity jet of steam hits a surface it suffers a change in momentum.
How does this give rise to a force on the surface?

Modern steam turbines produce power of the order of 1 000 000 kW with efficiencies up to about 47% (*Figure 3.7*).

*Figure 3.7   Steam turbines at NEI Parsons Heaton Works, Newcastle-upon-Tyne.*

## The internal combustion engine

Steam engines are external combustion engines in that the fuel is burnt externally, away from the engine where the movement is generated. In the internal combustion engine the fuel is burnt in the engine.

The first internal combustion engine was produced in 1833 by W.L. Wright but it was not until 1860 that the first commercial engine was produced by J.J.E. Lenoir. The driving force behind the development of the internal combustion engine was, however, Nikolaus August Otto. By 1868 he had developed an economic and commercially acceptable engine. The **four-stroke engine**, the form now used generally in cars,

*Figure 3.8 An 1898 Daimler car*

was produced by Otto in 1876. The first cars followed very shortly after this development (*Figure 3.8*).

---

**Question**   **9**   What would you consider to be the essential requirements of an engine that could be used to power a 'horseless carriage'?

---

*Figure 3.9* shows the basis of the four-stroke internal combustion engine. The down-stroke of the piston results in a mixture of air and petrol vapour being drawn into the cylinder. When the cylinder is full of vapour the valve through which the vapour enters the cylinder closes. The up-stroke of the piston compresses the air–vapour mixture. At a suitable compression the vapour is ignited by a spark produced at the sparking plug. The exploding mixture causes the piston to be pushed downwards. When the piston moves back up the cylinder the spent gases are expelled through a valve that opens. The entire cycle then repeats itself.

*Figure 3.10* shows how the pressure in the cylinder and the volume of the gases above the piston vary during the four-stroke cycle. This form of diagram is known as an **indicator diagram**. Work is done on the gas, i.e., energy is transferred to the gas, when it is compressed (*Figure 3.10(b)*). When the gas expands and pushes the piston down energy is transferred from the gas to the piston, i.e., work is done by the gas (*Figure 3.10(d)*). The shaded area in *Figure 3.10(f)* thus indicates the difference in the energy taken from the gas and the energy supplied to the gas. Thus, the larger this area the greater the energy supplied to the piston, and hence the car.

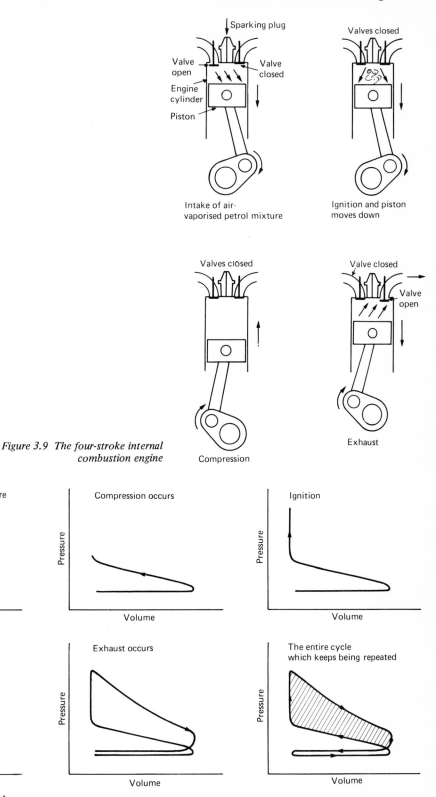

*Figure 3.9　The four-stroke internal combustion engine*

*Figure 3.10　The four-stroke cycle*

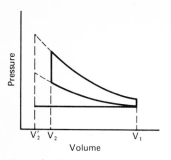

*Figure 3.11*

*Figure 3.11* shows an indicator diagram for a four-stroke cycle. The maximum volume of the air–vapour mixture in the cylinder is $V_1$, this being the space between the piston and the top of the cylinder when the piston is at the bottom of its stroke. When the mixture is compressed the volume at which ignition occurs, i.e., the volume above the piston at the top of its stroke, is $V_2$. The ratio $V_1/V_2$ is called the **compression ratio**. A higher compression ratio, e.g., $V_1/V_2'$, means a greater area enclosed by the pressure/volume graph and hence a greater energy output. Compression ratios between about 5 to 1 and 9 to 1 are generally used.

Of the total energy available from the combustion of the petrol only about 25% is actually used directly to propel the vehicle. The radiator and heat losses directly from the radiator account for about 50% of the energy and the remaining 25% is dissipated via the exhaust gases.

**Question   10**   Petrol of 97 octane value can produce about 43.7 MJ/kg. If the density of the petrol is 0.74 kg/litre about how much energy will be available for directly propelling the vehicle when one litre of petrol is used?

## The conversion of heat into work

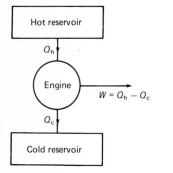

*Figure 3.12   The energy flow for a heat engine*

The steam engine, steam turbine and internal combustion engine are all methods of converting heat into work. All the engines can be represented by *Figure 3.12*. In all of them there is a heat input to the engine from some high temperature source, e.g., the boiler in the Newcomen engine. In all of them there is some wasted heat, e.g., that from a car engine. This waste heat always flows to a region of lower temperature, the radiator in the case of the car engine. The difference between the heat input $Q_h$ from the high temperature source and the waste heat $Q_c$ to the lower temperature region gives the energy that is useful, i.e., the work $W$.

$$W = Q_h - Q_c$$

The **thermal efficiency** of the engine is defined as the ratio of the useful work to the heat input to the engine. Thus

$$\text{Thermal efficiency} = \frac{W}{Q_h} = \frac{Q_h - Q_c}{Q_h} = 1 - \frac{Q_c}{Q_h}$$

The thermal efficiency only refers to the efficiency of the engine at converting heat to work and is not the overall efficiency of the engine concerned with converting the fuel from its 'raw' state into work.

The thermal efficiencies of all the engines considered in this chapter, and indeed engines in general, are low. None of the engines has a thermal efficiency of 100%, i.e., none of the engines convert all the heat input into work. An engine with an efficiency of 100% is not precluded by the conservation of energy principle – all the heat energy could according to this principle be converted into heat. But it never happens.

*Figure 3.13* shows how the measured thermal efficiencies of some reasonably modern power stations are related to the temperature of the

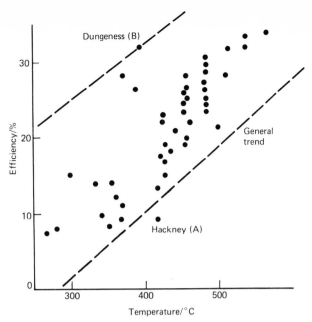

*Figure 3.13 Efficiency of power stations plotted against steam temperature*

steam. Though the newer power stations are more efficient than the older ones there does seem to be a relationship between efficiency and steam temperature. The higher the steam temperature the more efficient the power station. The efficiency is, however, by our equation above, related to the heat input to the engine from the hot reservoir and the heat output to the cold reservoir, i.e., there is a flow of heat through the engine from hot to cold. The flow of heat from a hot to a cold region, as for instance in conduction along a metal bar, depends on the temperature difference between the two regions. There would thus seem to be an implied relationship between $Q_c/Q_h$ and the temperatures of the hot and cold reservoirs. We can define a temperature scale such that

$$\frac{Q_c}{Q_h} = \frac{T_c}{T_h}$$

This is a definition of the **Kelvin temperature scale.** With temperatures measured in this way

$$\text{Thermal efficiency} = 1 - \frac{T_c}{T_h}$$

For an engine operating with steam at 400 °C, i.e., 673 K, and discharging heat into perhaps a cooling tower or river at 300 K then the thermal efficiency possible would be

$$\text{Thermal efficiency} = 1 - \frac{300}{673}$$

This is about 0.55 or 55%. The Dungeness power station operates with steam at about this temperature and has an efficiency of about 32%. The actual efficiency could thus be improved; it could be brought up to 55%. This is, however, the maximum efficiency possible – no more than 55% is possible for the temperatures being used.

**Questions**

11    Calculate the maximum thermal efficiency possible for an engine using steam at a temperature of 800 K and discharging heat into a reservoir at 300 K.

12    How would the answer to question **11** change if the cold reservoir was not maintained at 300 K but rose to a temperature of 350 K?

13    The Newcomen engine used steam at 100 °C and exhausted heat at probably a temperature of the order of 60 °C.
What would the maximum efficiency have been?

14    Watt modified the Newcomen engine by introducing a condenser external to the cylinder. This had the effect of reducing the temperature of the cold reservoir into which the heat was exhausted. Suppose the temperature of the cold reservoir was reduced to 30 °C — what effect would this have had on the possible efficiency?

---

The discussion in this section has been rather brief and superficial — it might be considered as just an introduction to a topic which is called the **second law of thermodynamics.** This could be stated as: it is impossible to convert heat energy entirely into work, some of the heat must always be wasted as heat.

---

**Suggestions for answers**

1    The other side of the piston is acted on by the atmospheric pressure and hence there is a pressure difference between the two sides of the piston.

2    $10^5 \times 0.55$ J or about 55 kJ

3    $46 \times 9.8 \times 46$ or about 21 kJ. Lower than the answer to question **2.** The pressure probably never fell to zero and so the pressure difference was less than $10^5$ N/m$^2$, also the piston probably never moved the entire length of the cylinder. If the correct data had been known the two answers should have been equal.

4    4.2 kW

5    308 kW

6    When the velocity changes there is a momentum change. A change in momentum means that a force is acting on the water. But a force acting on the water means that there must be an opposing and equal force acting on the hosepipe (Newton's third law).

7    The direction of the vector representing the change in momentum needs to lie in the plane of the turbine so that the reaction force causes the turbine to rotate (*see Figure 3.14*). You might like to refer to *Book 1: Motion and Force* if you are not sure of this answer. The direction of the momentum change is affected by the angle of the blades.

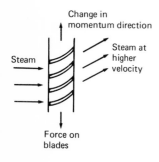

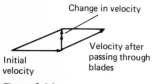

*Figure 3.14*

8     A change in momentum means a force. The force acting on the steam causing its momentum to change will be balanced by an opposite and equal force acting on the surface (Newton's third law). The force is equal to the rate of change of momentum.

9     Not too heavy. The fuel should be compact and light, giving a high energy output from a small volume and mass. Large enough power to propel the car.

10     25% of 43.7/0.74 MJ or about 15 MJ

11     62.5%

12     55.25%

13     About 15%

14     It would have increased the efficiency to about 19%.

---

**Further problems**     No suggestions for answers are given for these problems.

15     On the basis of a consideration of thermal efficiency explain why the use of high pressure steam was able to lead to an improvement in efficiency.

16     List the different forms of energy involved in a cycle of operation of the Newcomen engine.

17     What was the purpose of the separate condenser in the Watt engine?

18     What advantages did the steam engine have over the windmill as a source of power?

19     The advent of the steam engine was the beginning of the period in British history known as the industrial revolution. What other technological changes were taking place at the same time as the invention and development of the steam engine? You might like to read Chapter 3 in *The Industrial Revolution 1760–1830* by T.S. Ashton (Oxford University Press).

20     What is the thermal efficiency possible with a heat engine operating between temperatures of 200 °C and 27 °C?

21     What is the thermal efficiency of an engine that takes in 1000 J of heat and produces 150 J of work?

22     Explain how a four-stroke internal combustion engine operates.

# 4 Energy resources

*Lee Hall colliery and Rugely power station*

**Objectives**

The intention of this chapter is to consider energy in its economic, political and social contexts. It has been assumed that Chapters 1 and 2 of this book have been covered. The manipulation of numbers with indices is used quite a lot in this chapter and familiarity with this is assumed.

The general objectives for this chapter are that after working through it you should be able to:

(a) Discuss the way in which the world's use of energy is increasing, recognising some of the factors involved;
(b) Discuss the likely run-down of the world's supply of fossil fuels;
(c) Discuss the economic and social effects of the world's use of energy.

**Teaching note**

There are a considerable number of publications to which students can refer, e.g., *The Energy Question* by G. Foley, *Man and Energy* by A.R. Ubbelohde, *Energy and Power* – A Scientific American Book, *Man, Energy, Society* by E. Cook, *Coal, the Basis of Nineteenth-century Technology* – AST281 Unit 4 The Open University, *Energy Resources* – S266 Block 2 The Open University.

78

## 'Using' energy

Coal, oil and gas are examples of what we call fuels. A power station can burn a fuel and the resulting release of energy be used to drive a turbine. This in turn drives a generator and results in the production of electricity. The electrical energy can then be transmitted to your home where it can be used to supply heat or drive machines (*Figure 4.1*).

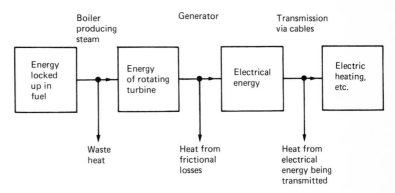

Figure 4.1

**Question    1**    Write down the energy transformation sequence for petrol being used to propel a car.

When a fuel is burnt we must still have energy conservation and no matter how many transformations the energy undergoes the total energy must remain constant. The energy does, however, become spread around. It would seem highly improbable that we could take all the various parcels of energy that were produced by the burning of a fuel and recon-centrate them to give us the concentrated source of energy that was the fuel. Thus when we talk of 'using' the energy from fuels we are really talking of the dissipation of energy that was initially concentrated.

## The world's energy requirements

It has been estimated that at the beginning of this century, in the one year 1900, the world used about $0.3 \times 10^{20}$ J. The estimate for the year 1950, halfway through the century, is about $0.7 \times 10^{20}$ J. It has been estimated that by the end of the century, the year 2000, the world will be using about $7 \times 10^{20}$ J per year. Thus in the first 50 years of this century the rate of energy usage per year more than doubled. In the second 50 years the rate of energy usage is expected to increase by

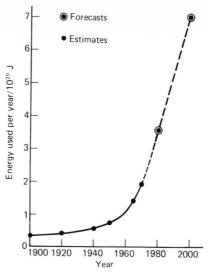

Figure 4.2   The world's use of energy

a factor of ten. *Figure 4.2* shows this type of change graphically and very clearly indicates the very large increase in the energy used per year.

---

**Question   2**   How would you obtain from *Figure 4.2* the total energy that has been used up to the year 1950? Up to the year 1970? The anticipated use up to the year 2000?

---

Why does the rate of use of energy increase as the years pass? There appear to be two factors: the increase in the world population and the increasing use of machines. In the year 1900 the population of the world was about $1.6 \times 10^9$, in 1950 about $2.3 \times 10^9$ and for the year 2000 the estimate is of $7 \times 10^9$. In 1900 the average energy used per person per year was

$$\frac{0.3 \times 10^{20}}{1.6 \times 10^9} \quad \text{or about } 19 \times 10^9 \text{ J}$$

In 1950 this average had become about $30 \times 10^9$ J and the anticipated figure for the year 2000 is $100 \times 10^9$ J.

The above figures are the averages per person when the entire world is considered. Thus, in the first 50 years of this century the average energy used per person increased by a factor of about one and a half, in the second 50 years the average is expected to increase by a further factor of about three.

For the United States the energy used per person per year was reasonably constant up to the year 1900. By 1950, however, the energy per person had increased by a factor of about three (compare this with the world change over that same time). By the year 2000 the energy used per person is expected to have increased by a factor of more than three times the 1950 figure, i.e., about 9 times the 1900 figure. This type of increase is characteristic of a nation becoming industrialised.

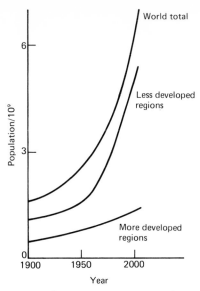

*Figure 4.3 World population growth*

The less developed countries, as they change to more industrialised societies, will certainly want to increase the amount of energy they use per person. A 'technological' man uses about ten times the amount of energy an 'advanced agricultural' man uses and twenty times that used by a 'primitive agricultural' man.

**Question** 3 *Figure 4.3* shows how the world population is changing in total and also the changes for both the less developed and more developed parts of the world.
How do you think these trends will affect the energy needs of the less developed and more developed parts of the world?

### The world's energy resources

With the exception of radioactivity, the tides due to the moon and the earth's geothermal energy, the sun is the provider of all the various forms of energy that exist on the earth. The earth receives energy from the sun of about $170 \times 10^{15}$ J every second. A significant amount of this energy is reflected or re-radiated back out into space, but some 60% is absorbed in the earth's atmosphere or at the surface. The radiation absorbed in the atmosphere and at the earth's surface provides heat energy which leads to convection currents within the atmosphere, i.e., winds, and within the oceans. Evaporation lifts water, via rain, to mountain tops and so results in water flowing into rivers. Thus the windmills and water-mills make use of energy that originated from the sun. The fossil fuels make use of the sun's energy that became 'locked up' in living matter. All living matter depends on the sun for energy and fossil fuels are nothing more than dead matter that once, long ago, lived.

**Question    4**    Estimate the energy needed to evaporate sufficient water to produce a cloud. A single cloud might contain sufficient water vapour
to produce a rainfall which gives 20 mm of rain over an area of
about 1 km$^2$.

The world stocks of the fossil fuels existing, and accessible to man,
have been estimated as:

| | |
|---|---|
| Coal | $2000 \times 10^{20}$ J |
| Oil | $120 \times 10^{20}$ J |
| Natural gas | $110 \times 10^{20}$ J |
| Tar sand and shale oil | $30 \times 10^{20}$ J |
| Total | $2260 \times 10^{20}$ J |

If you look in books quoting similar data you will no doubt find different figures — there are many different estimates of the amounts of fossil
fuels.

How long will these reserves of fossil fuels last? Present data, *Figure
4.2*, shows that we are using something like $2 \times 10^{20}$ J per year and that
this might rise to $7 \times 10^{20}$ J per year by the year 2000. At an average
rate of use of about $5 \times 10^{20}$ J per year then the fossil fuels might be
expected to last about 450 years. The fuels are not, however, each being
used at the same rate, and also it seems reasonable to expect that over
this longer period the rate of use will increase to a considerably higher
figure.

In the year 1860 coal was the only fossil fuel having any significant
use. Wood was, however, widely used. The following data shows some
of the general world trends.

| *Year* | *Energy produced per year*/$10^{18}$ J | | |
|---|---|---|---|
| | *Coal* | *Oil* | *Natural gas* |
| 1860 | 38 | 0 | 0 |
| 1900 | 208 | 8 | 3 |
| 1920 | 358 | 38 | 9 |
| 1940 | 421 | 112 | 31 |
| 1960 | 600 | 402 | 179 |

**Questions    5**    What percentage of the energy extracted from fossil fuels came
from coal in (a) 1900 and (b) 1960?

**6**    The first of the following extracts is taken from an advertisement
by the National Coal Board of Britain, the second from an article
at the same time, in the Daily Telegraph (19.7.77).

**A source of energy that will last for 300 years**
At the present rate of production, Britain has proved coal
reserves which will last at least 300 years. This puts Britain's
Coal Industry in a strong position alongside strictly limited oil
and gas supplies, and the continuing development of nuclear
power. With this assured energy supply, based on coal, British
Industry can plan ahead with confidence.

**Beauty vale pits plan must go ahead, Ezra insists**

A peaceful spread of English countryside will be transformed into a busy mining centre if the National Coal Board is allowed to proceed with plans announced yesterday. The board wants to open three collieries in the Vale of Belvoir, an unspoilt area of open farmland in the north-east corner of Leicestershire. In spite of protests from environmentalists, the board intends to pursue its case for working the 500 million tons of coal beneath the Vale. Sir Derek Ezra, board chairman, said yesterday: 'We are convinced that we must work the coal if we are to supply the nation's needs'. . . .

(a) Why do you think there has been a decrease in the percentage of the world's energy requirements that is being met by coal?

(b) The amount of coal being mined is still, however, increasing — what are the problems associated with this increase?

7    *Figure 4.4* shows how the rate of use of an exhaustible resource is likely to vary with time. The area under the graph represents the total amount of the resource. Plot the data given in the above table for oil, and with the information that the total stocks remaining after 1960 were about $360 \times 10^{20}$ J estimate when the oil is likely to run out. Your estimate will involve some 'inspired' guessing.

Figure 4.4

There are many estimates as to when the fossil fuels will run out. Coal seems likely to last of the order of 400 years, oil and gas perhaps only 40 years.

There are other sources of energy which do not arise from fossil fuels and thus can affect both the time for which fossil fuels will last and also perhaps determine the future of mankind when the fossil fuels are failing to supply the world's energy requirements. One very significant source of energy is nuclear energy. Heat is produced in a nuclear reactor and this can be used to produce steam and drive a turbine in the same way as heat produced from burning a fossil fuel (*Figure 4.5*). The first

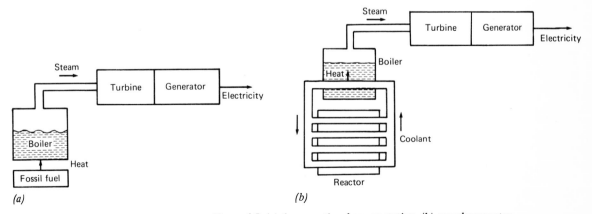

*Figure 4.5  (a) A conventional power station, (b) a nuclear reactor*

commercial nuclear power station was built in England at Calder Hall and first produced power in 1956. The following table shows the world trend occurring in the development of nuclear energy.

| Year | Nuclear energy per year/$10^{18}$ J |
|------|-------------------------------------|
| 1963 | 40 |
| 1966 | 56 |
| 1969 | 240 |
| 1972 | 500 |

It has been estimated that by the year 2000 about 20% of the world's energy may come from nuclear energy.

**Questions**    8    Use the above data and that in *Figure 4.2* to give an estimate of the percentage of the world energy produced by nuclear energy in 1972.

9    The following quotation is taken from an article by H. Ellis in *New Scientist* (29 April 1971, p. 272).

### No reactors in New York City

The fall from grace of science in the United States has been loudly proclaimed once again. Stimulated by citizen protest, the Atomic Safety and Licensing Board has refused permission for Columbia University to fuel and operate its already built Triga reactor. At best the situation is ironic; at worst it is tragic.

The irony is that when fission was new and reactor dynamics unknown, the first neutron-multiplying pile in history was built at Columbia. Now that the atom has developed a lily-white safety record for a quarter of a century, the public that used to take pride in nuclear technology will not tolerate the safest of research reactors in its midst. The tragedy is that New York City has both a power shortage and filthy air. The available solution is smokeless power from nuclear plants. But a city that will not tolerate a Triga is most unlikely to accept a power reactor.

(a) Why do you think the citizens do not want a reactor in the city?
(b) What do you think will be the problem facing the citizens if they do not have a nuclear reactor supplying power?

It has been estimated that there are vast reserves of uranium and thorium and that there would be no immediately foreseeable limit to the nuclear energy available. The limit, if there is one, might be imposed by the other materials used in the construction of the reactors.

A significant contribution to the world's energy supplies is made by hydroelectricity. This depends on water falling from a height and having its potential energy transformed to kinetic energy during the fall. This kinetic energy then causes a turbine to rotate and so produce electricity at a generator. In 1971, hydroelectricity provided about $4.7 \times 10^{18}$ J.

It has been estimated that this could be increased by a factor of about eight. The requirements for a hydroelectric power station are rivers which can be dammed to provide a harnessable fall for the water. One of the problems is the silt carried by rivers — this can decrease the storage capacity of the reservoir created behind the dam.

**Question 10** (a) What is the potential energy of 1 kg of water at a height of 100 m above a turbine?
(b) What is the kinetic energy of the 1 kg of water falling through the 100 m and striking the turbine blades?
(c) What mass of water has to hit a turbine per second if the turbine produces 50 MJ of electricity from the generator? Assume that 80% of the kinetic energy of the water is transformed into electricity.
(d) What does this tell you about the river flow into the reservoir?

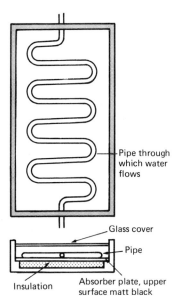

*Figure 4.6 A flat-plate solar energy collector*

Labels on figure: Pipe through which water flows; Glass cover; Pipe; Absorber plate, upper surface matt black; Insulation

Hydroelectricity is essentially a result of solar energy causing water to evaporate at low levels and become deposited at higher levels. Why not use solar energy directly? The earth receives about $3 \times 10^{24}$ J per year at its surface; this is considerably more than the present energy requirements per year. Although this is a large amount of energy it is spread very thinly over the earth's surface. At its maximum it does not amount to more than about 1 $kW/m^2$ in England. Thus, if all the energy over a square metre of surface in some spot in England could be gathered it would not amount, at its maximum, to more than that produced by a small electric fire. Over and entire year it only averages out at about 100 $W/m^2$ in England. Some other parts of the world do, however, have values as high as 250 $W/m^2$. Thus to tap any significant amount of the sun's energy would require collectors with very large areas. Hence use of direct solar energy does not appear to be the answer to the world's energy requirements when the supplies of fossil fuels run down. It might, however, be a cheap source of low power energy for some domestic purposes such as slightly raising the temperature of the water in a swimming pool or a summer supply of hot bath water.

**Questions 11** How great a surface area of solar collector would be needed, if 80% efficient and in England at its most sunny, to provide an energy output equivalent to that produced by a 100 MW power station?

**12** *Figure 4.6* shows the form of a typical flat plate solar collector that might be used for domestic purposes.

(a) Both the tube through which the water flows and the plate behind the tubing are painted matt black — why?
(b) Why is there insulation behind the backing plate?
(c) What do you think is the purpose of the backing plate?

How will our future energy supplies be met? Coal? Oil? Gas? Nuclear energy? Hydroelectricity? Solar energy collectors? Some other new method — perhaps energy from the tides? Some old method — perhaps energy from the wind? *Figure 4.7* shows a prediction up to the year 2000. Perhaps it will happen like that — perhaps not.

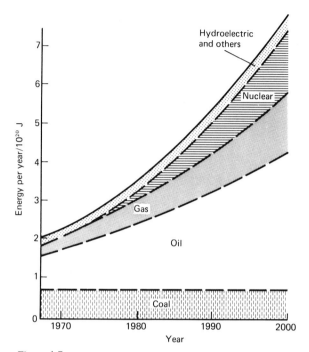

*Figure 4.7*

**Suggestions for answers**

1. Energy locked up in fuel, potential energy in a gas, kinetic energy of piston, kinetic energy of car, heat generated by friction between car and road. Also there is heat produced at each of the transformations.

2. The area under the graph of a rate against time gives the quantity with which the rate is concerned. Thus in the case of *Figure 4.2* the area under the graph is the total energy used. To determine the total energy up to 1950 requires you to consider the extrapolation of the graph back to the time when the energy used per year was effectively zero. You then need to estimate the area from this zero point up to the year 1950. As a very rough estimate I would suggest that up to 1900 the area might be $100 \times 10^{20}$ J and that from 1900 to 1950 it is about $20 \times 10^{20}$ J. This would give for the total energy used up to 1950 an estimate of $120 \times 10^{20}$ J. From 1950 to 1970 my estimate is about $20 \times 10^{20}$ J and thus a total up to 1970 of $140 \times 10^{20}$ J. From 1970 to 2000 my estimate is about $240 \times 10^{20}$ J and so a total up to 2000 of $380 \times 10^{20}$ J. If you are not sure about the area under a rate graph you might like to look at Chapter 1 in *Book 1:*

*Motion and Force* where a graph of velocity against time was considered. Velocity is the rate at which distance is covered and so the area under a velocity/time graph is distance.

3    The population in the less developed countries is increasing more rapidly than that in the more developed countries. They are also becoming more industrialised. This would suggest that there may be a considerable demand for more energy in the future.

4    The specific latent heat of vaporization of water is about $2 \times 10^6$ J/kg (*see* Chapter 2). The volume of water deposited by the cloud is $10^6 \times 20 \times 10^{-3} = 2 \times 10^4$ m$^3$. Water has a density of 1000 kg/m$^3$ and thus the mass of water in the cloud is about $2 \times 10^7$ kg. Hence the energy required is about $4 \times 10^{13}$ J.

5    (a) About 95%, (b) about 51%

6    (a) You might like to consider how coal-fired central heating compares with oil- or gas-fired central heating. The oil and gas are less messy, more easily made automatic, give less pollution. Some of the purposes for which coal was used have vanished, e.g., railway locomotion.
     (b) Coal mining considerably affects the area in which the mine is located, e.g., the presence of slag heaps. There is also the unpleasant job of extracting the coal from the pits.

7    *Figure 4.8(a)* is a plot of the data and shows how the energy used per year varies with time. The area under the graph gives the total energy used. Thus up to 1920 this amounted to about $400 \times 10^{18}$ J. Up to 1940 about $2000 \times 10^{18}$ J. Up to 1960 about $9000 \times 10^{18}$ J. The stock still remaining is about $36\,000 \times 10^{18}$ J. This means that the total initial stock was about $45\,000 \times 10^{18}$ J and we are thus about 20% of the way through the stocks. The peak is about 1990, according to my roughly drawn graph in *Figure 4.8(b)*.

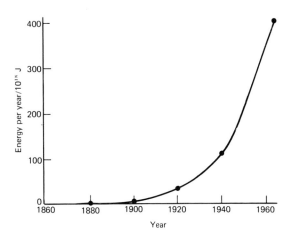

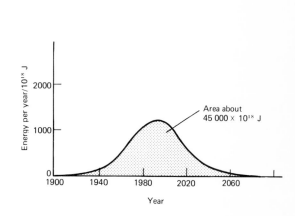

*Figure 4.8*

**8**    $[500 \times 10^{18}/(200 \times 10^{20})] \times 100\%$ or 2½%.

**9**    (a) Fear of a nuclear hazard.
(b) Where do they obtain power which does not pollute the environment?

**10**    (a) $1 \times 9.8 \times 100 = 980$ J
(b) 980 J
(c) $50 \times 10^6 = 0.80 \times m \times 9.8 \times 100$, hence $m$ is about $64 \times 10^3$ kg/s.
(d) The river flow must replenish the water flowing out of the reservoir if the level is not to drop. A hydroelectric power station will have more than one turbine.

**11**    Output of 80 W/m², hence an area of $100 \times 10^6 /80 = 1.25 \times 10^6$ m
or 1.25 km².

**12**    (a) To give maximum absorption of the radiation.
(b) So that the backing plate and the pipe reach the maximum temperature possible — the heat loss through the back of the collector being minimised.
(c) To absorb radiation and heat the pipe.

---

**Further problems**    No suggestions for answers are given for these problems.

**13**    The following data gives the world production of steel per year.

| Year | Steel production per year/$10^9$ kg |
|------|-------------------------------------|
| 1870 | 1 |
| 1900 | 29 |
| 1910 | 60 |
| 1920 | 72 |
| 1929 | 121 |
| 1940 | 142 |
| 1950 | 189 |
| 1960 | 380 |

(a) What would you estimate the world production of steel to be in the year 2000?
(b) It has been estimated that to produce 1 kg of finished steel requires about $48 \times 10^6$ J. Would any limitation in energy supplies be expected to have any significant effect on the production of steel?

**14**    It has been estimated that in about the year 1970 the numbers of cars per thousand of the population was as follows:

| Country | Cars per 1000 population |
|---------|-------------------------|
| Canada | 310 |
| France | 240 |
| Great Britain | 210 |
| Japan | 85 |
| USA | 430 |

(a) What effect will these numbers have on the energy requirements of the countries concerned?

(b) Petrol supplies about 40 MJ/kg. The average car will give about 10 km/kg.

What do you think is the average distance covered by a car in a year? Hence estimate the average energy requirements of a car per year.

(c) The overall average number of cars per thousand of population for the world is about 50. Estimate the world requirement for energy for cars.

15    It has been estimated that about 80% of the world's coal reserves are held by the USA and USSR.

What do you think would happen if these two nations collaborated to (a) force down, (b) force up the price of coal?

16    In 1970 a person in a developed country used about 5000 kg of coal (it may have been used on his behalf). In an undeveloped country the figure was about 400 kg.

(a) Why the difference?

(b) It has been estimated that in the year 2000 these figures will have changed to 11 000 kg and 1000 kg.

Do you consider these estimates to be reasonable?

17    The population of North America was, in 1964, about $2 \times 10^8$ and the energy used by them in that year was about $4 \times 10^9$ J.

(a) How much energy did the average person in North America use in that year?

(b) Use the data given in *Figure 4.2* and *Figure 4.3* to estimate the percentage of the world's energy used by North America in that year and how the average energy used per person in North America compared with the world average.

18    It is thought that vegetation has existed on earth for about $10^8$ years. The total amount of fossil fuels that has been produced in this time has been estimated as being capable of giving about $10^{23}$ J. At the present time the energy reaching the earth's surface from the sun amounts to about $3 \times 10^{24}$ J per year. Assuming that this has not changed over the years, estimate the percentage of this energy reaching the earth which is converted into fossil fuels.

19    Produce a design for a simple solar energy collector.

20    Consider the problems of developing wind power as a significant contribution to the energy needs of a country.

(a) Air has a density of about 1.2 kg/m$^3$. At a wind speed of 10 m/s what mass of air will hit sails of area 40 m$^2$ per second?

(b) What will be the kinetic energy transferred from the wind to the sails per second?

(c) If the windmill has an efficiency of about 50% what will be the power developed by the windmill?

(d) How many of such windmills would be needed to produce the energy supplied by a typical power station of 100 MW?

(e) What would be the problems associated with using windmills to generate a significant amount of the energy requirements of a country?

---

### Appendix 4.1    An energy problem

The following article by D.R. Cope and P. Hills appeared in *New Scientist*, 25 October 1979. It summarises the problems that are posed by a proposal to start a new group of coal mines (see question **6** which concerns the same problem).

**What the NCB wants to do**

After initial borehole exploration of the area in 1974, planning applications for coal extraction and surface developments of the proposed new coal field were submitted in July 1978. The Secretary of State for the Environment subsequently called these in.

The field covers an area of some 234 km² (91 square miles), with its centre about 20 km south-east of Nottingham. It lies under parts of Leicestershire, Nottinghamshire and Lincolnshire, although all the surface developments would be in Leicestershire. The coal extends well outside the actual Vale of Belvoir and two of the proposed mine sites — Asfordby and Saltby — are on the edge of the field and not within the vale.

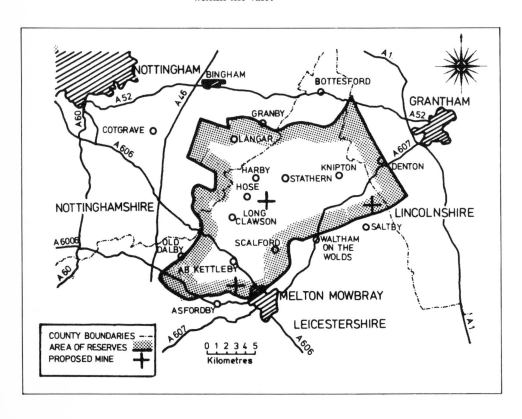

The Vale of Belvoir itself is a flat, low-lying area, largely devoted to mixed farming, with limited woodland and open space. However, to the south of the vale lies an attractive marlstone escarpment — the Harby Hills — which extends eastward to Belvoir Castle, home of the Duke of Rutland. The escarpment bears extensive woodland.

The vale, especially the western part, also has a commuter population largely working in Nottingham, while throughout there are small industrial firms, many serving the area's farming activities. The nearest existing mine is at Cotgrave, Nottinghamshire, about 3 km from the north-west edge of the new coal field.

The coal measures in the area, which are geologically part of the Yorkshire, Derbyshire and Nottinghamshire coal field, contain five 'target' seams at depths ranging from 400 to 850 m, with an average depth of 600 m and thicknesses of 1.2 to 2.1 m. Total workable reserves are estimated at 510 million tonnes, with one seam, the Deep Main, supplying 160 million tonnes of this. The five seams do not all overlap in a workable configuration in any part of the coal field; 26 per cent of the area contains one workable seam, 32 per cent two seams, 33 per cent three seams and 9 per cent four seams.

After a detailed investigation of possible approaches to winning the coal, the NCB opted for three 'total' mines (with each site handling men, materials and coal) at Hose, Asfordby and Saltby. The coal measures are overlain by heavily waterlogged Bunter Sandstones. These and other geological conditions mean that access to the coal will have to be by shafts rather than inclined drifts.

The proposal envisages an annual output of 7 million tonnes of saleable coal — 3 million tonnes from Hose and 2 million tonnes each from the other sites. This would give Belvoir a working life of some 70–80 years.

The three mines' output will be removed by rail, and some new spur lines to existing routes will have to be constructed. The declared destinations of the coal are the Trent-side CEGB power stations. Most of the output of Nottinghamshire, Leicestershire and Derbyshire pits is burnt at these stations and the new coal field is seen as a replacement for the exhausting capacity in these areas, particularly north-west Leicestershire. However, the report by the NCB's consultants acknowledges the possibility of other markets 'in the longer term'.

### Mountains of waste

Along with the saleable coal, the three pits will produce about 2.7 million tonnes of mining waste a year. This is some 30 per cent of the total 'run-of-mine' and is an unavoidable result of the geology of the coal seams. The proposals include provision for tipping of the waste near each of the three mines, using derelict land wherever possible. In all, there is an estimated land need of 682 hectares for a 50-year tipping period. This compares with a total of 100 ha for the pithead sites themselves. Remote disposal of the waste was investigated by the NCB's consultants, the most attractive site being the disused Fletton clay brickwork pits of Bedfordshire, about 130 km away. This option was rejected as 'extremely expensive' and as raising its own environmental problems. Underground disposal of waste has been ruled out as too costly, deleterious to underground working conditions and safety, and, at best, able to handle only a small proportion of the waste produced. Coal preparation procedures intended for the new pits would use filter presses to remove water from tailings and so eliminate the large settlement lagoons.

Subsidence is not expected to be as severe as in areas of shallower workings. The absence of old workings should also minimise subsidence. Recently published NCB estimates are that 35 per cent of the coal field area would subside by less than 1 m, 40 per cent by 1–2 m, 20 per cent by 2–3 m, and 5 per cent by more than 3 m. Statistical extrapolations from existing subsidence impacts in other areas suggest that about 30 per cent of the 3500 conventional houses in the coal field might be affected by subsidence. Of these, about 64 per cent would have 'very slight', 32 per cent 'slight', 3 per cent 'appreciable' and 1 per cent 'severe' damage, according to NCB calculations. This excludes dwellings such as terraced cottages which require special consideration, as well as industrial and historic buildings. The NCB estimates that 1.5 per cent of the 207 km$^2$ of agricultural land in the area could be 'adversely' affected by subsidence.

Present estimates suggest that about 3800 men would be required to operate the Belvoir pits, with a further 300 administrative staff. Ancillary jobs which could be associated with the development bring the estimated total number of jobs created to about 6000 and the total associated population to 18 000–20 000. NCB statements refer to some of the workforce coming from declining pits in north-

*Saltby mine site*

west Leicestershire and other parts of the Nottinghamshire/Derbyshire coal field, as well as the local area.

The workforce is expected to occupy about 5000 houses; exact location of these would depend on the private/public tenure split and the commuting pattern that might develop. Most new housing would probably be on the outskirts of Melton Mowbray, Grantham and Bingham, where existing local plans envisage expansion. The construction workforce is expected to peak at around 1700 in the middle of the mine-building period. The Hose and Asfordby mines will be developed simultaneously, with Saltby following four years later. Plans are for the entire project to be fully operational within 13 years of the starting date.

## Acknowledgements

The publishers would like to acknowledge the following for kindly supplying illustrations and extracts and/or permission to reproduce them.

British Leyland – for the illustration on page 65

Education Development Center Inc – for *Figure 1.2* from *P.S.S.C. Physics*, (1965) D.C. Heath & Co; Lexington, Ma

Professor A. Keller – for the original of the cover illustration. (From A. Keller and A. O'Connor (1958) *Discuss. Faraday Soc.*, **25**, 114)

National Coal Board – for the illustration on page 78

North of Scotland Hydro-Electric Board – for the illustration on page 6

Resora Ltd – for the illustration on page 35

Reyrolle Parsons Ltd – for *Figure 3.7*

Shell U.K. Limited – for the illustration on page 1

Science Museum, London – for *Figures 3.4* and *3.8*, Crown copyright reserved


Addison-Wesley – for the extract from *Feynman Lectures on Physics* Volume 1 by R.P. Feynman, R.B. Leighton and M. Sands (1963)

*Daily Telegraph* – for the extract on page 83

National Coal Board – for the extract on page 82

*New Scientist,* London, the weekly review of science and technology – for the article by D.R. Cope and P. Hills in *Appendix 4.1* and the extract on page 84

Oxford University Press – for the extract on page 32 from *The Second Law* by H.A. Bent

Penguin Books – for the extract on page 33 from *The Laws of Physics* by M.A. Rothman


## Bibliography

*Coal, the Basis of Nineteenth-Century Technology*, AST 281 Unit 4, The Open University; Milton Keynes (1973)

*Energy and Power*, A Scientific American Book, W.H. Freeman and Co.; Reading (1972)

*The Energy Question* by G. Foley, Penguin; Harmondsworth (1976)

*Energy Resources*, S266 Block 2, The Open University; Milton Keynes (1973)

*Heat Engines* by J.F. Sandfort *Science Study Series No. 22* Heinemann Educational; London (1964)

*Man and Energy* by A.R. Ubbelohde, Penguin; Harmondsworth (1963)

*Man, Energy and Society* by E. Cook, W.H. Freeman and Co.; Reading (1976)

*Nuffield Physics: Pupils' Text Year 4* Longmans/Penguin; Harmondsworth (1978)

*Physical Science Study Committee Laboratory Guide for Physics* D.C. Heath; Lexington, Ma. (1965)

*Physics Investigations* by W. Bolton, Wheaton; Exeter (1976)

*Project Physics Course, Handbook* by F.J. Rutherford *et al.*, Holt, Rinehart and Winston; New York (1970)

*Project Physics Course, Text* by F.J. Rutherford *et al.*, Holt, Rinehart and Winston; New York (1970)

*Project Technology Handbook No. 11: Industrial Archaeology of Watermills and Waterpower* Heinemann Educational; London (1970)

*A Simple History of the Steam Engine* by J.D. Storey, Baker; London (1969)

# Index